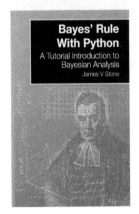

James V Stone, Honorary Associate Professor, University of Sheffield, UK.

# A Short Introduction to the Mathematics of Information Theory

## James V Stone

Title: A Short Introduction to the Mathematics of
Information Theory
Author: James V Stone

©2022 Sebtel Press

First Edition, 2022.
Typeset in LaTeX $2_\varepsilon$.
First printing.

ISBN 9781739672720

Cover image Claude Shannon (1916-2001). Courtesy MIT Museum.

*For Talaska*

The founding fathers of the modern computer age - Alan Turing, John von Neumann, Claude Shannon - all understood the central importance of information theory, and today we have come to realise that almost everything can either be thought of or expressed in this paradigm. I believe that, one day, information will come to be viewed as being as fundamental as energy and matter.

Demis Hassabis (CEO and co-founder of DeepMind) 2017.

# Contents

# Preface

**Who Should Read This Book?** Even though this book is fairly brief for a technical subject, it retains the tutorial style of writing adopted for more extended books in this Tutorial Introduction series. Consequently, this book act as a primer for students and researchers who wish to gain a firm grasp of the basics of information theory.

As Mark Twain famously said, "I didn't have time to write a short letter, so I wrote a long one instead". Similarly, it is relatively easy to write a long book, which appeases every tangential whim of thought with a circuitous route to a handful of rarefied facts perched on some distant horizon. But writing a short book, which discards all but the most essential facts, is much, much harder (involving removal of much, much repetition). So here it is; the essence of information theory in about 70 pages.

**Corrections.** Please email corrections to j.v.stone@sheffield.ac.uk. A list of corrections can be found at

`https://jamesstone.sites.sheffield.ac.uk/books/infoshort`.

**Acknowledgements**. There are no acknowledgements. This is all my own work, and no bastard lifted a single finger to help.

Jim Stone, Sheffield, England, 2022.

# Chapter 1

# The Laws of Information

JVS PUBLISH THIS AS A HARDBACK ONLY.

## 1.1. Why Information Theory?

Let's begin with a really simple example. Imagine you use your computer to send an email which says, "Meet me at 8, and don't be late." You type the letters m, e, e, and t, and then you glance at the screen to see the word "neet". Either you have hit the wrong key, or the computer made an error, so there are at least two possible explanations.

The first explanation is that your brain intended to type m. However, the message from your brain to your finger had to travel through several feet of nerve fibres before it reached your finger, and the message got corrupted along the way by a small amount of random noise. This noise altered the message just enough for your finger to hit the wrong key.

The second explanation is you have hit the correct key. However, the message from the keyboard to the computer had to travel through several feet of wire before it reached the computer, and the message got corrupted along the way by a small amount of random noise. This

1

noise altered the message just enough for the computer to display the wrong letter.

Whichever explanation is correct, in both cases, the problem is that the message got corrupted by noise as it travelled from the sender (i.e. your brain or the keyboard) to the receiver (your finger, or the computer). Indeed, the problem of noise lies at the heart of information theory. An obvious solution is to simply remove the noise. Unfortunately, noise is everywhere. There is no measurement you can make, no signal you can receive, that does not contain some un-wanted components in the form of noise. When you de-tune a radio, the hiss you hear is white noise, and when you tune into a radio station, the noise does not disappear, it just becomes less audible.

Of course, noise is not restricted to radios, brains and computer keyboards. In fact, it applies to Wifi systems, satellite communication, TV signals, speech communication, and the ability to perceive the world through eyes, ears and fingers.

So, what has all this got to do with information theory? Well, in essence, information theory tells us exactly how much information we can communicate as the amount of noise varies. Most remarkable of all, information theory guarantees that it is often possible to communicate

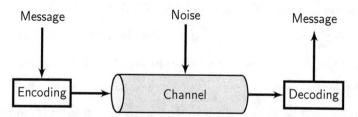

Figure 1.1: The communication channel. A message (data) is encoded before being used as input to a communication channel, which adds noise. The channel output is decoded by a receiver to recover the message.

without error, and that it is always possible to communicate with a vanishingly small amount of error, even in the presence of noise. Consequently, when a signal is sent from a satellite to a TV satellite dish, or from a radio station to a radio, or from an eye to the brain, information theory tells us how it is possible for a signal to be received with only a vanishingly small amount of error.

> **Key point**. Information theory guarantees that it is always possible to communicate with (at most) a vanishingly small amount of error, even in the presence of noise.

## 1.2. Information Theory

The universe is conventionally described in terms of physical quantities such as mass and velocity, but a quantity at least as important as these is *information*. Whether we consider computers[15], evolution[2;9], physics[7], black holes[13], artificial intelligence[5], quantum

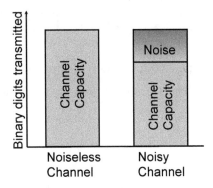

Figure 1.2: The *channel capacity* of noiseless and noisy channels is the maximum rate at which information can be communicated. If a noiseless channel communicates data at 10 binary digits/s then its capacity is $C = 10$ bits/s. The capacity of a noiseless channel is numerically equal to the rate at which it communicates binary digits, whereas the capacity of a noisy channel is less than this because it is limited by the amount of noise in the channel.

3

computation[22], or the brain[8;21], we are driven inexorably to the conclusion that the behaviour of these systems is determined mainly by the way they process information.

In 1948, Claude Shannon published a paper called *A Mathematical Theory of Communication*[23]. This paper heralded a transformation in our understanding of information. Before Shannon's paper, information had been viewed as a kind of poorly defined miasmic fluid. But after Shannon's paper, it became apparent that information is a well-defined and, above all, *measurable* quantity.

## 1.3. The Laws of Information

The theory of information theory is defined in terms of a few key *mathematical theorems*, where a theorem is just a mathematical statement which has been proven to be true. Information theory underpins so many applications and fields of research that these theorems deserve to be called the *laws of information*.

These laws can be summarised with reference to Figure 1.1. Don't worry if you only have an informal understanding of the terms used here; they will be explained more fully in later chapters.

A *source* (e.g. a theatre stage) generates *messages* (e.g. actors speaking) with a fixed amount of information per second. These messages are *encoded* (e.g. by a TV camera) before being *transmitted* through a *communication channel* (satellite) with a fixed *capacity*, and the channel output is *decoded* (e.g. by a TV) to form either, a) the original messages if the information rate from the source is less than the channel capacity, or, b) an approximation to those messages if the information rate from the source is more than the channel capacity.

Given these definitions, an informal summary of the the laws of information is as follows. For any communication channel:

1. There is a limit, the *channel capacity*, to the rate at which information can be transmitted through a channel.

2. The *source coding theorem* states that the channel capacity can be reached by judicious packaging, or *encoding*, of input data.

3. The *noisy channel coding theorem* states that the channel capacity shrinks in proportion to the amount of *noise* in the channel.

4. The *rate distortion theorem* states that, if the amount of information in the input exceeds the channel capacity then output values can still be used to estimate input values, but the uncertainty in those estimates cannot be less than the difference between the input information and the channel capacity.

In order to understand these laws properly, it is necessary to have a rigorous definition of exactly what is meant by the terms ɪuncertainty and *amount of information*, which will be defined in the next chapter. For the present, it is sufficient to know only two facts. First, a standard unit of information is the *bit*, which provides enough information to choose between two equally probable alternatives. Second, just as a pint bottle can contain up to one pint of liquid, so a binary digit can represent up to one bit of information.

## 1.4. Encoding Images

Suppose we wanted to transmit an image of $100 \times 100$ pixels, in which each pixel has more than two possible grey-level values. A reasonable number of grey-level values turns out to be 256, as shown in Figure 1.3a.

(a)                                                    (b)

Figure 1.3: Grey-level image. (a) An image in which each pixel has one out of 256 possible grey-levels, between 0 and 255, each of which can be represented by a binary number with 8 binary digits (e.g. 255=11111111). (b) Histogram of grey-levels in the picture.

There are large regions of the image that look as if they contain only one value. In fact, each such region contains values which are similar, but not identical, as shown in Figure 1.4. The similarity between adjacent pixel values means that the value of one pixel can be predicted (to some extent) by the values of nearby pixels. This, in turn, means that adjacent pixel values are not *independent* of each other, and that the image has a degree of *redundancy*. How can this observation be used to encode the image?

One method consists of encoding the image in terms of the differences between the grey-levels of adjacent pixels. For brevity, we will call this *difference coding*. In principle, pixel differences could be measured in any direction within the image, but, for simplicity, we concatenate consecutive rows to form a single row of 10,000 pixels, and then take the difference between adjacent grey-levels. We can see the result of difference coding by 'un-concatenating' the rows to reconstitute an image, as shown in Figure 1.5a, which looks like a badly printed version of Figure 1.3a.

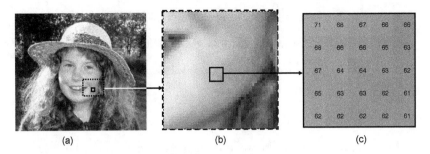

Figure 1.4: Adjacent pixels tend to have similar grey-levels, so the image has a large amount of redundancy, which can be used for efficient encoding. (a) Grey-level image.  (b) Magnified square from a.  (c) Magnified square from b, with individual pixel grey-levels indicated.

If adjacent pixel values are similar then the difference between the values is close to zero. In fact, a histogram of difference values shown in Figure 1.5b shows that the most common difference values are indeed close to zero, and only rarely greater than ±63. Thus, using difference coding, we could represent almost every one of the 9,999 difference values in Figure 1.5a as a number between −63 and +63.

In those rare cases where the difference is larger than ±63, we could list these separately as each pixel's location (row and column as 2×7 binary digits), and its grey-level value (8 binary digits). Most coding procedures have special 'housekeeping' fragments of computer code to deal with things like this, which incurs a negligible extra cost.

At first, it is not obvious how difference coding represents any saving over simply sending the value of each pixel's value. However, because these differences are between −63 and +63, they span a range of 127 different values, i.e. $[-63, -62, \ldots, 0, \ldots, 62, 63]$. Any number in this range can be represented using seven binary digits, because $7 = \log 128$ (which leaves one spare value).

In contrast, if we were to send each pixel's value in Figure 1.3a individually, then we would need to send 10,000 value. Because each value could be anywhere between 0 and 255, we would have to send eight binary digits ($8 = \log 256$) for each pixel.

Once we have encoded an image into 9,999 pixel difference values $(d_1, d_2, \ldots, d_{9999})$, how do we use them to reconstruct the original image? Well, if the difference $d_1$ between the first pixel value $x_1$ and the second pixel value $x_2$ is, say, $d_1 = (x_2 - x_1) = 10$ value and the value of $x_1$ is 5, then we obtain the original value of $x_2$ by adding 10 to $x_1$; that is, $x_2 = x_1 + d_1$ so $x_2 = 5 + 10 = 15$. We then continue this process for the third pixel ($x_3 = x_2 + d_2$), and so on. Thus, provided we know the value of the first pixel in the original image (which can be encoded as eight binary digits), we can use the pixel value differences to recover the value of every pixel in the original image. The fact that we can reconstruct the original image (Figure 1.3a) from the pixel value differences (Figure 1.5a) proves that they both contain exactly the same amount of *information*.

Let's work out the total saving from using this difference coding method. The naive method of sending all 10,000 pixel values, which vary between 0 and 255, would need eight binary digits per pixel, requiring a total of 80,000 binary digits. But using difference coding we would need seven binary digits per difference value, making a total of only 70,000 binary digits.

As we shall see in subsequent chapters, a histogram of data values (e.g. image grey-levels) can be used to obtain an upper bound for the average amount of information each data value could convey. It turns out that the histogram (Figure 1.3b) of the grey-levels in Figure 1.3a

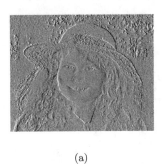

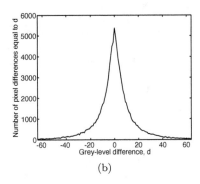

(a)                                      (b)

Figure 1.5: Difference coding. (a) Each pixel grey-level is the difference between adjacent horizontal grey-level values in Figure 1.3a (grey = zero difference). (b) Histogram of grey-level differences between adjacent pixel grey-levels in Figure 1.3a. Only differences between ±63 are plotted.

yields an upper bound of 7.84 bits/pixel. In contrast, the histogram (Figure 1.5b) of the grey-level differences in Figure 1.5a yields an upper bound of just 5.92 bits/pixel.

Given that the images in Figures 1.3a and 1.5a contain the same amount of information, and that Figure 1.5a contains no more than 5.92 bits/pixel, it follows that Figure 1.3a cannot contain more than 5.92 bits/pixel either. This matters because Shannon's source coding theorem guarantees that if each pixel's grey-level contains an average of 5.92 bits of information then (in principle) we should be able to represent Figure 1.3a using no more than 5.92 binary digits per pixel. But the estimate of 5.92 bits/pixel still represents an upper bound. In fact, the smallest number of binary digits required to represent each pixel is numerically equal to the amount of information (measured in *bits*) implicit in each pixel. So if we want to know the smallest number of binary digits that could be used to represent each pixel grey-level then what really want to know is: how much information does each pixel contain?

This is a hard question, but we can get an idea of the answer by comparing the amount of computer memory required to represent the image in two different contexts.

First, to display the image on a computer screen, the value of each pixel requires eight binary digits of computer memory, so the bigger the picture, the more memory it requires to be displayed.

Second, a compressed version of the image can be stored on the computer's hard drive using fewer than eight binary digits per pixel (e.g. using difference coding). Consequently, storing the (compressed) version of an image on the hard drive requires less memory than displaying that image on the screen.

For example, if the image in Figure 1.3a is has 10,000 pixels, where each pixel's grey-level value is between 0 and 255, then each pixel can be represented on a computer screen as eight binary digits (because $2^8 = 256$), or one *byte*. However, when the file containing this image is inspected, it is typically found to contain only 50,000 bytes; in other words, the image in Figure 1.3a can be compressed by a factor of $2(= 10,000/50,000)$. This means that the information implicit in each pixel, which requires eight binary digits for it to be displayed on a screen, can be represented with four binary digits on a hard drive.

This is important because it implies that each set of eight binary digits used to display each pixel of Figure 1.3a contains an average of only four bits of information; and therefore each binary digit contains an average of only *half a bit* of information. At first sight, this seems like an odd statement. But, as we shall see in Section 3.4, a fraction of a bit is a well-defined quantity, with a reasonably intuitive interpretation.

## 1.5. A Short History of Information Theory

Even the most gifted scientist cannot command an original theory out of thin air. Just as Einstein could not have devised his theories of relativity if he had no knowledge of Newton's work, so Shannon could not have created information theory if he had no knowledge of the work of Boltzmann (1875) and Gibbs (1902) on thermodynamic entropy, Wiener (1927) on signal processing, Nyquist (1928) on sampling theory, or Hartley (1928) on information transmission[19].

Shannon was not alone in trying to solve one of the key scientific problems of his time (i.e. how to define and measure information), but he was alone in being able to produce a complete mathematical theory of information: a theory that might otherwise have taken decades to construct. In effect, Shannon single-handedly accelerated the rate of scientific progress, and it is entirely possible that, without his contribution, we would still be treating information as if it were some ill-defined vital fluid.

# Chapter 2

# Defining Information, Bit by Bit

## 2.1. Shannon's Desiderata

Now that we have a little experience of information, we can consider how it ought to behave in more general contexts. Shannon knew that in order for a mathematical definition of information to be useful it had to meet a particular minimal set of conditions.

1. **Additivity**. The total information of a set of individual outcomes (e.g. coin flips) is obtained by adding the information of those individual outcomes.

2. **Continuity**. The information of an outcome (e.g. obtaining a head after flipping a coin) increases continuously (i.e. smoothly) as the probability of that outcome decreases.

3. **Symmetry**. The information of a sequence of outcomes does not depend on the order in which those outcomes occur.

Crucially, Shannon[25] proved that the definition of information given below is the only one which meets all of these conditions. We begin by exploring additivity.

## 2.2. Decisions, Decisions

Information is usually measured in *bits*, and one bit is the amount of information required to choose between two equally probable, or *equiprobable*, alternatives.

To understand why this is true, imagine you are standing at the fork in the road at point A in Figure 2.1, and that you want to get to the point marked D. The fork at A represents two equiprobable alternatives, so if I tell you to go left then you have received one bit of information. If we represent my instruction with a *binary digit* (0=left and 1=right) then this binary digit provides you with one bit of information which tells you which road to choose.

> **Key point.** A bit is the *amount of information* required to choose between two equiprobable alternatives.

Now imagine that you come to another fork, at point B in Figure 2.1. Again, a binary digit (1=right) provides one bit of information, allowing you to choose the correct road, which leads to C. Note that C is one of four possible interim destinations that you could have reached after making two decisions. The two binary digits that allow you to make the correct decisions provided two bits of information, allowing you to choose from four (equiprobable) alternatives; 4 equals $2 \times 2 = 2^2$.

A third binary digit (1=right) provides you with one more bit of information, which allows you to again choose the correct road, leading to the point marked D. There are now eight roads you could have chosen from when you started at A, so three binary digits (which provide you with three bits of information) allow you to choose from eight equiprobable alternatives; 8 equals $2 \times 2 \times 2 = 2^3$.

We can restate this in more general terms if we use $n$ to represent the number of forks and $m$ to represent the number of final destinations. If you have come to $n$ forks then you have effectively chosen from $m = 2^n$ final destinations. Because the decision at each fork requires one bit of information, $n$ forks require $n$ bits of information.

Viewed from another perspective, if there are $m = 8$ possible destinations then the number of forks is $n = 3$, which is the *logarithm* of 8. Thus, $3 = \log_2 8$ is the number of forks implied by eight destinations. More generally, the logarithm of $m$ is the power to which 2 must be raised in order to obtain $m$; that is, $m = 2^n$. Equivalently, given a number $m$ which we wish to express as a logarithm, $n = \log_2 m$. The subscript $_2$ indicates that we are using logs to the base 2 (all logarithms in this book use base 2 unless stated otherwise).

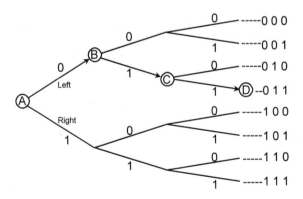

Figure 2.1: For a traveller who does not know the way, each fork in the road requires one bit of information to make a correct decision. The 0s and 1s on the right-hand side summarise the instructions needed to arrive at each destination; a left turn is indicated by a 0 and a right turn by a 1.

If we simply double the number of destinations then we would need $\log 2m$ bits, which can be written as

$$\log 2m \quad = \quad \log m + \log 2 \qquad (2.1)$$

$$= \quad n + 1 \text{ bits.} \qquad (2.2)$$

In other words, if we need $n$ bits to navigate to one out of $m$ equiprobable destinations then we need $n + 1$ bits to navigate to one out of $2m$ equiprobable destinations.

> **Key point.** If you have $n$ bits of information then you can choose from $m = 2^n$ equiprobable alternatives. Equivalently, if you have to choose between $m$ equiprobable alternatives then you need $n = \log_2 m$ bits of information. If you have to choose between $2m$ equiprobable alternatives then you need $n+1$ bits.

## 2.3. Bits Are Not Binary Digits

The word *bit* is derived from *binary digit*, but they are fundamentally different types of quantities. A binary digit is the value of a binary variable, where this value can only be a 0 or a 1. In contrast, a bit is an *amount of information*, and the number of bits conveyed by a binary digit (when averaged over both of its states) can vary between zero and one. In the example above, the amount of information conveyed by each binary digit was chosen to be exactly 1 bit, but that is not always the case, as we shall see below. By analogy, just as a pint bottle can carry between zero and one pint of liquid, so a binary digit can convey between zero and one bit of information.

> **Key point.** A bit is the *amount of information* required to
> choose between two equiprobable alternatives (e.g. left/right),
> whereas a binary digit is the *value of a binary variable*, which
> can adopt one of two possible values (i.e. 0/1).

## 2.4. Surprisal

Just as there were 2 equiprobable outcomes at each fork in the road
in the example above, so there are 2 equiprobable outcomes when a
coin is flipped. And, just as 1 bit is required to choose the correct road
from the $m = 2$ alternatives at each fork, so 1 bit is required to predict
which of $m = 2$ equiprobable outcomes will be obtained from a single
coin flip. Conversely, observing the outcome of a single coin flip with
$m = 2$ equiprobable outcomes provides 1 bit, whether the outcome is
a head $x_h$ or a tail $x_t$. The amount of information associated with a
single event is called the *surprisal*, represented with a lower case $h$ here

$$h(x_h) = \log m = \log 2 = 1 \text{ bit} \tag{2.3}$$

$$h(x_t) = \log m = \log 2 = 1 \text{ bit}. \tag{2.4}$$

We can adopt a different but equivalent perspective, which seems
almost trivial in the case of a fair coin, but which yields insight for
unfair coins, as we shall see. For a fair coin, the probability of a head is
$p(x_h) = 1/m = 0.5$, so that the amount of information obtained when
a head is observed is given by its surprisal

$$h(x_h) = \log m = \log 1/p(x_h) = \log 1/0.5 = 1 \text{ bit}, \tag{2.5}$$

and similarly, the surprisal of a tail $h(x_t) = 1$ bit.

Now consider an unfair coin, which is so bent out of shape that it lands heads up 90% of the time. When this coin is flipped, we expect it to land heads up, so when it does so we are less surprised than when it lands tails up. The more improbable a particular outcome is, the more surprised we are to observe it. Accordingly, the amount of Shannon information or the *surprisal* of a head is

$$h(x_h) \quad = \quad \log \frac{1}{p(x_h)} \text{ bits,} \qquad (2.6)$$

which is often expressed as

$$h(x_h) \quad = \quad -\log p(x_h) \text{ bits.} \qquad (2.7)$$

For the bent coin with $p(x_h) = 0.9$, the surprisal comes to a mere $h(x_t) = 0.15$ bits (we shall return to this example below). Note that Shannon did not pluck the definition of surprisal out of thin air. Instead, it turns out that this definition is the only one that meets the conditions specified in Section 2.1.

> **Key Point.** The Shannon information is a measure of surprise.

## Shannon Entropy is Average Surprisal

In practice, we are not usually interested in the surprisal of a particular value of a random variable, but we are interested in how much surprisal, on average, is associated with the entire set of possible values that the variable can adopt. The average surprisal of a random variable $X$ is called its entropy, which is defined by the *probability distribution* of

$X$. For a coin, because the variable $X$ can adopt just two values, the probability distribution of a coin also consists of just two values, the probability of a head $p(x_h)$ and the probability of a tail $p(x_t)$, so the probability distribution of a coin is

$$p(X) \quad = \quad \{p(x_h), p(x_t)\}, \qquad (2.8)$$

and the entropy of the distribution $p(X)$ is represented as $H(X)$.

> **Key Point.** The entropy of a random variable is the average surprisal associated with each possible value of that variable.

**A Note on Nomenclature**

The terms surprisal, information and Shannon entropy are well defined, but they are often used interchangeably in the literature on information theory. Indeed, they are occasionally used interchangeably in this text, where the intended meaning is fairly obvious from the context, as

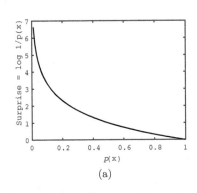
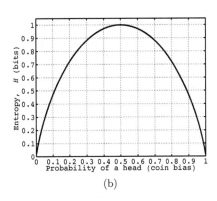

(a)                                             (b)

Figure 2.2: (a) Shannon information as surprise. Values of $x$ that are less probable have larger values of surprise, defined as $\log_2(1/p(x))$ bits. (b) Graph of entropy $H(X)$ versus coin bias (probability $p(x_h)$ of a head). The entropy of a coin is the average amount of surprise or Shannon information in the distribution of possible outcomes (i.e. heads and tails).

2 Defining Information, Bit by Bit

explained in Section 2.8. Incidentally, the term Shannon entropy is often used to differentiate it from the term entropy used in physics [27].

## 2.5. The Entropy of Flipping a Coin

### The Entropy of a Fair Coin

The average amount of surprisal associated with the outcome of a coin flip can be found as follows. If a coin is fair or unbiased then $p(x_h) = p(x_t) = 0.5$; so the Shannon information gained when a head or a tail is observed is the surprisal $\log 1/0.5 = 1$ bit. If the surprisal of a head is 1 bit, and the surprisal of a tail is also 1 bit then the average Shannon information gained after each coin flip is the average surprisal, which is also 1 bit. Because entropy is defined as average surprisal, the entropy of a fair coin is therefore $H(X) = 1$ bit.

### The Entropy of an Unfair Coin

If a coin is biased such that the probability of a head is $p(x_h) = 0.9$ then it is easy to predict the result of each coin flip (i.e. with 90% accuracy if we predict a head for each flip). If the outcome is a head then the amount of Shannon information gained is $\log(1/0.9) = 0.15$ bits. But if the outcome is a tail then the amount of Shannon information gained is $\log(1/0.1) = 3.32$ bits. Notice that more information is associated with the more surprising outcome. Given that the proportion of flips that yield a head is $p(x_h)$, and that the proportion of flips that yield a tail is $p(x_t)$ (where $p(x_h) + p(x_t) = 1$), the average surprise is

$$H(x) \quad = \quad p(x_h) \log \frac{1}{p(x_h)} + p(x_t) \log \frac{1}{p(x_t)} \qquad (2.9)$$

$$= \quad (0.9 \times 0.15) + (0.1 \times 3.32) \qquad (2.10)$$

$$= \quad 0.469 \text{ bits}, \qquad (2.11)$$

as in Figure 2.2b. If we define a tail as $x_1 = x_t$ and a head as $x_2 = x_h$ then Equation 2.9 can be written as

$$H(X) \quad = \quad \sum_{i=1}^{2} p(x_i) \log \frac{1}{p(x_i)} \text{bits}. \qquad (2.12)$$

## 2.6. The Definition of Entropy

In accord with the equation for the entropy of a coin in Equation 2.12, a random variable $X$ with a probability distribution is $p(X) = \{p(x_1), \ldots, p(x_m)\}$ has an entropy of

$$H(X) \quad = \quad \sum_{i=1}^{m} p(x_i) \log \frac{1}{p(x_i)} \qquad (2.13)$$

$$= \quad E[\log 1/p(x_i)] \text{ bits}. \qquad (2.14)$$

This definition matters because Shannon's source coding theorem (see Section 5) guarantees that each value of the variable $X$ can be represented with an average of (just over) $H(X)$ binary digits.

## 2.7. Dicing with Entropy

### The Entropy of an 8-Sided Die

Throwing an 8-sided die (Figure 2.3a) yields an *outcome*, so there are 8 equiprobable outcomes. If we define an *outcome value* as the number on the upper face then there are $m = 8$ possible outcome values $A_x = \{1, 2, 3, 4, 5, 6, 7, 8\}$, represented by the symbols $x_1, \ldots, x_8$. Because all 8 outcomes are equally probable, it follows that the probability $P$ of each outcome value is $P = 1/8$, as shown in the normalised histogram in Figure 2.3b.

Using Equation 2.13, these 8 probabilities yield the entropy

$$
\begin{aligned}
H(x) &= \sum_{i=1}^{8} p(x_i) \log \frac{1}{p(x_i)} \quad\quad\quad\quad\quad (2.15) \\
&= p(x_1) \log \frac{1}{p(x_1)} + p(x_2) \log \frac{1}{p(x_2)} + \cdots + p(x_8) \log \frac{1}{p(x_8)}
\end{aligned}
$$

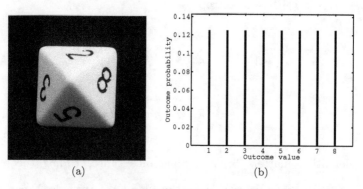

(a)　　　　　　　　　　　　(b)

Figure 2.3: (a) An 8-sided die. (b) The uniform normalised histogram (probability distribution) of outcomes has an entropy of $\log 8 = 3$ bits.

where $\log(8) = 3$, so that

$$H(x) \quad = \quad 1/8 \sum_{i=1}^{8} 3 \qquad (2.16)$$

$$= \quad 1/8 \times 8 \times 3 \qquad (2.17)$$

$$= \quad 3 \text{ bits.} \qquad (2.18)$$

In words, each die outcome provides 3 bits of information. Because the entropy is the average information provided by each outcome, and because each outcome provides the same amount of information in this case, the entropy of a single die is also 3 bits.

### The Entropy of a 6-Sided Die

Throwing a six-sided die yields an *outcome*, so there are 6 equiprobable outcomes. If we define an *outcome value* as the number on the upper face then there are $m = 6$ possible outcome values $A_x = \{1, 2, 3, 4, 5, 6\}$, represented by the symbols $x_1, \ldots, x_6$. Because all 6 outcomes are equally probable, it follows that the probability $P$ of each outcome value is $P = 1/6 = 0.1667$.

Using Equation 2.13, these 6 probabilities yield the entropy

$$H(x) \quad = \quad \sum_{i=1}^{6} p(x_i) \log \frac{1}{p(x_i)}, \qquad (2.19)$$

where $\log(1/0.1667) = \log(6) = 2.585$, so that

$$H(x) \quad = \quad 1/6 \sum_{i=1}^{6} 2.585 \qquad (2.20)$$

$$= \quad 1/6 \times 6 \times 2.585 \qquad (2.21)$$

$$= \quad 2.585 \text{ bits.} \qquad (2.22)$$

In words, each die outcome provides 2.585 bits of information. Because the entropy is the average information provided by each outcome, and because each outcome provides the same amount of information in this case, the entropy of a single die is also 2.585 bits.

**The Entropy of a Pair of Dice**

Throwing a pair of six-sided dice yields an *outcome* in the form of an ordered pair of numbers, and there are a total of 36 equiprobable outcomes, as shown in Table 2.1. If we define an *outcome value* as the sum of this pair of numbers then there are $m = 11$ possible outcome values $A_x = \{2, 3, 4, 5, 6, 7, 8, 9, 10, 11, 12\}$, represented by the symbols $x_1, \ldots, x_{11}$. Dividing the frequency of each outcome value by 36 yields the probability $P$ of each outcome value, as shown in Figure 2.4b and Table 2.1. Using Equation 2.13, we can use these 11 probabilities to find the entropy

$$
\begin{aligned}
H(x) &= p(x_1) \log \frac{1}{p(x_1)} + p(x_2) \log \frac{1}{p(x_2)} + \cdots + p(x_{11}) \log \frac{1}{p(x_{11})} \\
&= \sum_{i=1}^{11} p(x_i) \log \frac{1}{p(x_i)} = 3.27 \text{ bits.} \quad (2.23)
\end{aligned}
$$

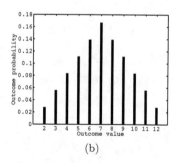

(a)                               (b)

Figure 2.4: (a) A pair of dice.    (b) Probability distribution (normalised histogram) of dice outcome values.

## 2.8. Interpreting Entropy

We already know that, if the entropy of a coin is $H(X) = 1$ bit then the variable $X$ can be used to represent $m = 2^1 = 2$ equiprobable values. By analogy, a variable with an entropy of $H(X)$ bits can be used to represent

$$m = 2^{H(X)} \text{ equiprobable outcome values.} \qquad (2.24)$$

For the 8-sided die considered above, the entropy of the distribution of outcome values is $H(X) = 3$ bits, which could therefore represent $2^3 = 8$ equiprobable values; this is consistent with the fact that the die has 8 sides. Similarly, for a 6-sided die, its entropy is $H(X) = 2.585$ bits, which could therefore represent $2^{2.585} = 6$ equiprobable values; and this is consistent with the fact that the die has 6 sides.

In the case to the two dice considered above, the entropy of the distribution of outcome values is $H(X) = 3.27$ bits, which could

| Symbol | Sum | Outcome | Frequency | $P$ | Surprisal |
|---|---|---|---|---|---|
| $x_1$ | 2 | 1:1 | 1 | 0.03 | 5.17 |
| $x_2$ | 3 | 1:2, 2:1 | 2 | 0.06 | 4.17 |
| $x_3$ | 4 | 1:3, 3:1, 2:2 | 3 | 0.08 | 3.59 |
| $x_4$ | 5 | 2:3, 3:2, 1:4, 4:1 | 4 | 0.11 | 3.17 |
| $x_5$ | 6 | 2:4, 4:2, 1:5, 5:1, 3:3 | 5 | 0.14 | 2.85 |
| $x_6$ | 7 | 3:4, 4:3, 2:5, 5:2, 1:6, 6:1 | 6 | 0.17 | 2.59 |
| $x_7$ | 8 | 3:5, 5:3, 2:6, 6:2, 4:4 | 5 | 0.14 | 2.85 |
| $x_8$ | 9 | 3:6, 6:3, 4:5, 5:4 | 4 | 0.11 | 3.17 |
| $x_9$ | 10 | 4:6, 6:4, 5:5 | 3 | 0.08 | 3.59 |
| $x_{10}$ | 11 | 5:6, 6:5 | 2 | 0.06 | 4.17 |
| $x_{11}$ | 12 | 6:6 | 1 | 0.03 | 5.17 |

Table 2.1: A pair of dice has 36 possible outcomes.
Sum: outcome value, total number of dots for a given throw of the dice.
Outcome: ordered pair of dice numbers that could generate each symbol.
Freq: number of different outcomes that could generate each outcome value.
$P$: the probability that the pair of dice yields a given outcome value (freq/36).
Surprisal: $P \log(1/P)$ bits.

therefore represent $2^{3.27} = 9.65$ equiprobable values; as if we had a die with 9.65 sides. For the bent coin, where $H(X) = 0.469$ bits (Equation 2.11), the variable $X$ could be used to represent $m = 2^{0.469}$ or 1.38 equiprobable values; as if we had a die with 1.38 sides. Therefore, a variable with an entropy of $H(X)$ bits provides enough information to choose between $m=2^{H(X)}$ equiprobable alternatives.

At first sight, these statements seem strange. Nevertheless, translating entropy into an equivalent number of equiprobable alternatives serves as an intuitive guide for the amount of information represented by a variable.

> **Key Point.** Entropy is average information. A variable with an entropy of $H(X)$ bits provides enough information to choose between $m=2^{H(X)}$ equiprobable alternatives.

**Entropy: Give and Take**

Entropy is a measure of *uncertainty*. When our uncertainty is reduced, we gain information, so information and entropy are two sides of the same coin. However, information has a rather subtle interpretation, which can easily lead to confusion.

Average information shares the same definition as entropy, but whether we call a given quantity information or entropy usually depends on whether it is being given to us or taken away. For example, if a variable has high entropy then our initial uncertainty about its value is large and is, by definition, exactly equal to its entropy. If we are told the value of that variable then, on average, we have been given an amount of information equal to the uncertainty (entropy) we initially had about its value. Thus, receiving an amount of information is equivalent to having exactly the same amount of entropy taken away.

# Chapter 3

# Differential Entropy

## 3.1. The Trouble With Entropy

So far, we have considered entropy in the context of discrete random variables (e.g. coin flipping). However, we also need a definition of entropy for continuous random variables (e.g. temperature). In many research fields, results obtained with discrete variables can easily be extended to continuous variables, but information theory is not one of those cases.

Fortunately, there are ways to generalise the discrete definition of entropy to obtain sensible measures of entropy for continuous variables [14;17;18;20;30].

To estimate the entropy of any variable, it is necessary to know the probability associated with each of its possible values. For continuous variables this amounts to knowing its *probability density function* or *pdf*, which we refer to here as its distribution. We can use a pdf as a starting point for estimating the entropy of a continuous variable by making a histogram of a large number of measured values. However, this reveals a fundamental problem, as we shall see below.

In order to make a histogram of any continuous quantity $X$, such as human height, we need to define the width $\Delta x$ of bins in the histogram. We then categorise each measured value of $X$ into one histogram bin, as in Figure 3.1. Then the probability that a randomly chosen value of $X$ is in a given bin is simply the proportion of values of $X$ in that bin. The entropy of this histogram is then given by the average surprisal of its bins (here indexed with $i$),

$$H(X^{\Delta}) = \sum_i (\text{prob } X \text{ is in } i\text{th bin}) \times \log \frac{1}{\text{prob } X \text{ is in } i\text{th bin}}, \quad (3.1)$$

where $X^{\Delta}$ indicates that we are dealing with a continuous variable which has been discretised using a histogram in which each bin has a width equal to $\Delta x$. We have purposely not specified the number of bins; in principle, it can be infinite. In practice, it suffices to simply use enough bins to include all of the values in our data set, as in Figure 3.1.

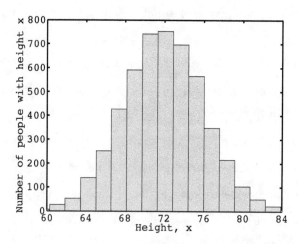

Figure 3.1: A histogram of $n = 5,000$ hypothetical values of human height $X$, measured in inches. This histogram was constructed by dividing values of $X$ into a number of intervals or bins, where each bin has width $\Delta x$, and then counting how many values are in each bin.

The probability that a randomly chosen value of $X$ is in the $i$th bin is given by the area $a_i$ of the $i$th bin, expressed as a proportion of the total area $A$ of all bins. A bin which contains $n_i$ values has an area equal to its height $n_i$ times its width $\Delta x$, $a_i = n_i \times \Delta x$, so that, if the sum of all bin areas is $A = \sum_i a_i$ then the probability that $X$ is in the $i$th bin is equal to the proportion of area occupied by the $i$th column $P_i = a_i/A$. It will prove useful later to note that the sum of these proportions (i.e. total area) of this *normalised histogram* is necessarily equal to 1:

$$\sum_i P_i = 1. \tag{3.2}$$

If the $i$th bin in this normalised histogram has height $p(x_i)$ and width $\Delta x$ then its area (height times width) can be obtained from

$$P_i = p(x_i) \, \Delta x, \tag{3.3}$$

which allows us to rewrite Equation 3.1 more succinctly as

$$H(X^\Delta) = \sum_i P_i \log \frac{1}{P_i}. \tag{3.4}$$

Given that the probability $P_i$ corresponds to the area of the $i$th column, we can interpret the height $p(x_i) = P_i/\Delta x$ of the $i$th column as a *probability density*. Substituting Equation 3.3 in Equation 3.4 yields

$$H(X^\Delta) = \sum_i p(x_i)\Delta x \times \log \frac{1}{p(x_i)\Delta x}. \tag{3.5}$$

However, given that the final term can be written as

$$\log \frac{1}{p(x_i)\,\Delta x} = \log \frac{1}{p(x_i)} + \log \frac{1}{\Delta x}, \qquad (3.6)$$

we can rewrite Equation 3.5 as

$$H(X^{\Delta}) = \left[\sum_i p(x_i)\,\Delta x \log \frac{1}{p(x_i)}\right] + \left[\log \frac{1}{\Delta x}\sum_i P_i\right],$$

where, according to Equation 3.2, the sum $\sum P_i = 1$, so that

$$H(X^{\Delta}) = \left[\sum_i p(x_i)\,\Delta x \log \frac{1}{p(x_i)}\right] + \log \frac{1}{\Delta x}. \qquad (3.7)$$

Thus, as the bin width approaches zero, the first term on the right becomes an integral, but the second term diverges to infinity:

$$H(X) = \left[\int_{x=-\infty}^{\infty} p(x)\log \frac{1}{p(x)}\,dx\right] + \infty. \qquad (3.8)$$

And there's the problem. For a continuous variable, as the bin width $\Delta x$ approaches zero, so $1/\Delta x$, and therefore $\log(1/\Delta x)$, and therefore the entropy of $X$, diverges to infinity.

One consequence of this is that the entropy of a continuous variable increases with the precision of our measurements (which determines the bin width). This makes sense if we bear in mind that increasing the precision of the measurements ought to increase the information associated with each measurement. For example, being told that a table is measured as five feet wide and that the device used to measure its width had a precision of ±0.1 inch provides more information than being told that the measurement device had a precision of ±1 inch. In practice, it means that we must always take account of the bin width when comparing the entropies of two different continuous variables. As we shall see, the problem of infinities disappears for quantities (e.g. mutual information) which involve the difference between entropies.

---

**Key point.** The estimated entropy $H(X^\Delta)$ of a (discretised) continuous variable increases as the width of bins in that variable's histogram decreases.

---

## 3.2. Differential Entropy

Equation 3.8 states that the entropy of a continuous variable is infinite, which is true, but not very helpful. If all continuous variables have infinite entropy then distributions that are obviously different have the same (infinite) entropy.

A measure of entropy called the *differential entropy* of a continuous variable ignores this infinity; it is defined as

$$H_{dif}(X) \quad = \quad \int_{x=-\infty}^{\infty} p(x) \log \frac{1}{p(x)} \, dx, \qquad (3.9)$$

where the subscript *dif* denotes differential entropy (although this is not used where the intended meaning is unambiguous). Thus, the differential entropy is that part of the entropy which includes only the 'interesting' part of Equation 3.8.

> **Key point**. The entropy of a continuous variable is infinite because it includes a constant term which is infinite. If we ignore this term then we obtain the *differential entropy* $H_{dif}(X) = \mathrm{E}[\log(1/p(x))]$, the mean value of $\log(1/p(X))$.

## 3.3. Interpreting Differential Entropy

The entropy of a continuous variable is a peculiar concept, inasmuch as it appears to have no meaning when considered in isolation. The fact that a variable has a definite amount of entropy tells us almost nothing of interest. In particular, knowing the amount of entropy of a variable does not place any limit on how much information that variable can convey (because each value of every continuous variable can convey an infinite amount of information). This stands in stark contrast to the case for a discrete variable, where entropy determines precisely how much information it can convey.

However, given that the accuracy of every measurement is limited by noise, this measurement noise places a strict upper limit on the information-carrying capacity of all continuous variables. Thus, even though each value of a continuous variable can, in principle, convey infinite information, the amount of information it conveys in practice depends on the accuracy of our measurements.

In effect, measurement noise divides up the range of probabilities of a continuous variable into a finite number of discrete intervals; the number of intervals increases as the measurement noise decreases. The exact consequences of this discretisation of continuous variables by measurement noise will be examined in Section 7.3.

> **Key point.** Noise limits the amount of information conveyed by a continuous variable and, to all intents and purposes, transforms it into a discrete variable with a finite number of discriminable values.

## 3.4. What is Half a Bit of Information?

If a variable has a uniform distribution then one bit halves our range of uncertainty about its value, just as it halves the number of possible routes in Figure 2.1. However, we often encounter fractions of a bit. So, what does it mean to have, say, half a bit of information?

We can find out what a fraction of a bit means by copying the recipe we use for whole bits (Equation 2.24). For clarity, we assume that the variable $X$ has a uniform distribution and that we know nothing about which value $X$ has. For example, the distribution of the 8-sided die shown in Figure 2.3b has an entropy of three bits, which corresponds to an initial uncertainty range of 1 or 100%. If we are given $H = 2$ bits of information about the value of $X$ then this reduces our range of uncertainty by a factor of $2^H = 2^2 = 4$, so our uncertainty about the value of a variable is one quarter as big as it was before receiving these two bits. Because the die has eight sides, this would mean that we now know the outcome is one of only two possible values (e.g. 3 or 8 ). More generally, if we treat our initial uncertainty as 1 (or 100%) then this implies that our *residual uncertainty* is $U = 1/4$ (or 25%), as shown in Figure 3.2. Note that the term residual uncertainty is unique to this book.

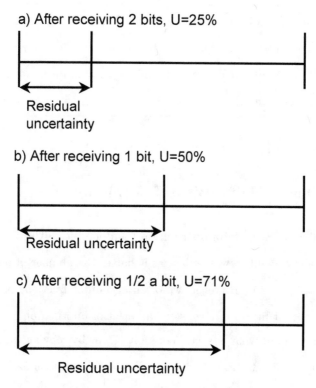

a) After receiving 2 bits, U=25%

Residual
uncertainty

b) After receiving 1 bit, U=50%

Residual uncertainty

c) After receiving 1/2 a bit, U=71%

Residual uncertainty

Figure 3.2: Residual uncertainty $U$. If we have no information about the location of a point on a line of length 1 then our initial uncertainty is $U = 1$. (a) After receiving 2 bits, we know which quarter contains the point, but we do not know where it is within that quarter, so $U = 1/4$. (b) After receiving 1 bit, we know which half contains the point, so our residual uncertainty is $U = 1/2$. (c) After receiving 1/2 a bit, we know the point lies within a region containing 0.71 of the line, so our residual uncertainty is $U = 0.71$.

Initially, our uncertainty spanned the whole line, which has a length of one. After receiving two bits, the region depicting the residual uncertainty has a length of 0.25. The precise location of this region depends on the particular information received, just as the particular set of remaining destinations in the navigation example in Section 2.2 depends on the information received.

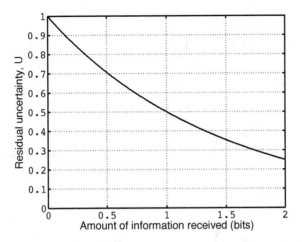

Figure 3.3: Residual uncertainty $U$ after receiving different amounts of information $H$, where $U = 2^{-H}$.

The recipe we have just used applies to any number of bits $H$, for which the residual uncertainty is

$$U = 2^{-H}, \qquad\qquad (3.10)$$

as shown in Figure 3.3. So, if we have $H = 1$ bit then our residual uncertainty is $U = 2^{-1} = 1/2$, which means that our residual uncertainty is half as big as it was before receiving this one bit.

Equation 3.10 applies to any value of $H$, including fractional values. It follows that if we receive half a bit ($H = 1/2$) then our residual uncertainty is $U = 2^{-1/2} = 0.71$, times as big as it was before receiving half a bit.

Let's keep this half a bit of information, and let's call our residual uncertainty $U_1 = 0.71$. If we are given another half a bit then our new residual uncertainty, which we call $U_2$, is our current uncertainty $U_1$ reduced by a factor of 0.71. Thus, after being given two half bits, our new residual uncertainty is $U_2 = 0.71 \times U_1 = 0.5$. So, as we should expect, being given two half bits yields the same residual uncertainty (0.5) as being given one bit.

> **Key point.** Complete uncertainty of $U=1.0$ is reduced to $U=2^{-0.5}=0.71$ after receiving half a bit of information. In general, uncertainty is $U=2^{-H}$ after receiving $H$ bits.

# Chapter 4

# Maximum Entropy Distributions

## 4.1. Why Maximum Entropy?

A distribution of values that has as much entropy (information) as theoretically possible is a *maximum entropy distribution*. Maximum entropy distributions are important because if we wish to use a variable to transmit as much information as possible then we had better make sure it has maximum entropy. For a given variable, the precise form of its maximum entropy distribution depends on the constraints placed on the values of that variable[20]. It will prove useful to summarise three important maximum entropy distributions. These are listed in order of decreasing numbers of constraints below.

The reason we are interested in maximum entropy distributions is because entropy equates to information, so a maximum entropy distribution is also a *maximum information distribution*. In other words, the amount of information conveyed by each value from a maximum entropy distribution is as large as it can possibly be. This matters because if we have some quantity $S$ with a particular distribution $p(S)$ and we wish to transmit $S$ through a communication channel, then we had better transform (encode) it into another variable $X$ with a maximum entropy distribution $p(X)$ before transmitting it.

Specifically, given a variable $S$, which we wish to transmit along a communication channel by encoding $S$ as another variable $X$, what distribution should $X$ have to ensure each transmitted value of $X$ conveys as much information as possible? For example, if $S$ is the outcome of throwing a pair of dice then the distribution of $S$ is shown in Figure 2.4b, which is clearly not uniform. More importantly, if we simply encode the outcome values $S$ between 2 and 12 as their corresponding binary numbers $X$, then the distribution of 0s and 1s in the resultant set of codewords is far from uniform. However, if $S$ is encoded as a binary variable $X$ (using Huffman coding, for example) then the distribution of 0s and 1s in the resultant set of codewords is almost uniform (i.e. the proportion of 0s and 1s is about the same). Of all the possible distributions of 0s and 1s, the uniform distribution has maximum entropy, and is therefore a *maximum entropy distribution*. Thus, Huffman coding implicitly encodes iid data as a maximum entropy distribution, which is consistent with the fact that it provides almost one bit per binary digit (i.e. it provides a fairly efficient code).

In contrast, for continuous variables, the distribution with maximum entropy is not necessarily uniform. We consider three types of continuous variable, each of which has a different particular constraint but is free to vary in every other respect. These constraints are:

1. fixed upper and lower bounds;

2. fixed mean, with all values greater than or equal to zero;

3. fixed variance (e.g. power).

Each of these constraints is associated with a different maximum entropy distribution. For the constraints above, the maximum entropy distributions are (1) uniform, (2) exponential, and (3) Gaussian.

## 4.2. The Uniform Distribution

Consider a random variable $X$ with fixed upper and lower bounds, distributed uniformly between zero and $a$, so the probability density $p(x)$ has the same value for all values of $X$, as in Figure 4.1. The width times the height of $p(X)$ must be one, so $p(x)\times a=1$, and so the probability density function of a uniform distribution is

$$p(x) = 1/a. \tag{4.1}$$

The probability density $p(x)$ of $x$ is therefore equal to $1/a$ between zero and $a$, and equal to zero elsewhere. By convention, a random variable $X$ with a uniform distribution which is non-zero between zero and $a$ is written as $X \sim U(0, a)$. The entropy of this uniform distribution is

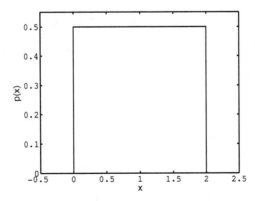

Figure 4.1: A uniform distribution with a range between zero and two has an area of one ($= 2 \times 0.5$), and an entropy of $\log 2 = 1$ bit.

therefore

$$H_{dif}(X) = \int_{x=0}^{a} p(x) \log a \, dx \qquad (4.2)$$

$$= \log a \text{ bits.} \qquad (4.3)$$

This result is intuitively consistent with the entropy of discrete variables. For example, in the case of a discrete variable, we know from Equation 2.2 that doubling the number of possible outcome values increases the entropy of that variable by one bit (i.e. by $\log 2$ bits). Similarly, for a continuous variable, doubling the range of continuous values effectively doubles the number of possible outcome values (provided we are content to accept that this number is infinitely large for a continuous variable) and also increases the entropy of that continuous variable by one bit. Thus, if the range of $X$ values is increased from $a$ to $b = 2a$ then the entropy of $Y = 2X$ should increase by exactly one bit in relation to the entropy of $X$, i.e.

$$H_{dif}(Y) = \log b = \log 2a = \log a + 1 \text{ bits.} \qquad (4.4)$$

More importantly, if a variable has a fixed lower and upper bound (e.g. zero and a) then *no probability distribution can have a larger entropy than the uniform distribution*[20].

> **Key point.** Given a continuous variable $X$ with a fixed range (e.g. between zero and two), the distribution with maximum entropy is the uniform distribution.

An odd feature of the entropy of continuous distributions is that they can have zero or *negative entropy*. For example, if $X$ has a range of $a = 1$ then $H_{dif}(X) = 0$, and if $a = 0.5$ then $H_{dif}(X) = -1$. One way to think about this is to interpret the entropy of a uniform distribution relative to the entropy of a distribution with an entropy of $H_{dif}(X) = 0$ (i.e. with a range of $a = 1$). If $a = 2$ then this distribution has an entropy which is $H_{dif}(X) = 1$ bit larger than the entropy of a distribution with $a = 1$. And if $a = 0.5$ then the distribution has an entropy which is one bit smaller than that of a distribution with $a = 1$. Similar remarks apply to the entropy of any continuous distribution.

## 4.3. The Exponential Distribution

An exponential distribution is defined by one parameter, which is its *mean*, $\mu$. The probability density function of a variable with an exponential distribution is

$$ p(x) \quad = \quad \begin{cases} \frac{1}{\mu} e^{-\frac{x}{\mu}} & x \geq 0 \\ 0 & x < 0, \end{cases} \tag{4.5} $$

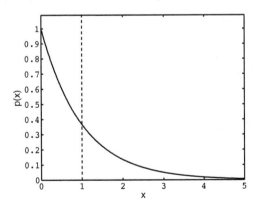

Figure 4.2: An exponential distribution with a mean of $\mu = 1$ (indicated by the vertical dashed line) has an entropy of $H_{dif}(X) = 1.44$ bits.

as shown in Figure 4.2 for $\mu = 1$. By convention, a random variable $X$ with an exponential distribution which has a mean of $\mu$ is written as $X \sim \exp(\mu)$. The entropy of an exponential distribution is [20]

$$H_{dif}(X) \quad = \quad \log e\mu \text{ bits.} \tag{4.6}$$

More importantly, if we know nothing about a variable except that it is positive and that its mean value is $\mu$ then the distribution of $X$ which has maximum entropy is the exponential distribution.

> **Key point.** Given a continuous positive variable $X$ which has a mean $\mu$, but is otherwise unconstrained, the distribution with maximum entropy is the exponential distribution.

## 4.4. The Gaussian Distribution

A Gaussian distribution is defined by two parameters, its *mean* $\mu$ and its *variance* $v$, which is the square of its *standard deviation* $\sigma$, so $v = \sigma^2$ (see Figure 4.3). The probability density function of a variable with a

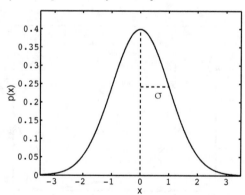

Figure 4.3: A Gaussian distribution with a mean of $\mu = 0$ and a standard deviation of $\sigma = 1$ (indicated by the horizontal dashed line), has an entropy of $H_{dif} = 2.05$ bits.

Gaussian distribution is

$$p(x) \quad = \quad \frac{1}{\sigma\sqrt{2\pi}}e^{-\frac{(x-\mu)^2}{2\sigma^2}}, \tag{4.7}$$

where the mean determines the location of the peak of the probability distribution, and the variance, which is the average squared difference between $x$ and the mean, $v = \mathrm{E}[(x-\mu)^2]$, determines how spread out the Gaussian distribution is. By convention, a random variable $X$ with a Gaussian distribution which has mean $\mu$ and variance $v$ is written as $X \sim N(\mu, v)$. The entropy of an iid Gaussian variable is [20]

$$
\begin{aligned}
H_{dif}(X) \quad &= \quad 1/2 \log 2\pi e \sigma^2 & (4.8) \\
&= \quad 1/2 \log 2\pi e + \log \sigma & (4.9) \\
&= \quad 2.05 + \log \sigma \text{ bits.} & (4.10)
\end{aligned}
$$

Given that $\log 1 = 0$, a Gaussian distribution with a standard deviation of $\sigma=1$ has an entropy of 2.05 bits. If $X$ is constrained to have a fixed variance $\sigma^2$ then *no probability distribution has larger entropy than the Gaussian distribution* [20].

> **Key point.** Given a continuous variable $X$ which has a variance $\sigma^2$, but is otherwise unconstrained, the distribution with maximum entropy is the Gaussian distribution.

# Chapter 5

# Noiseless Channels

Our objective is to transmit a message reliably through a communication channel shown in Figure 5.1. The message or *source signal s* is essentially a physical quantity, such as a sound or an image. The source signal is encoded as the channel input $x$. After the encoded message has been transmitted through the channel, the receiver observes the channel output $y$. The *channel capacity C* is the maximum amount of information that a channel can provide at its output about the input.

> **Key point.** *Channel capacity C* is the maximum amount of information that a channel output can provide about its input value, usually expressed in bits per second.

## 5.1. The Source Coding Theorem

Shannon's source coding theorem, described below, applies only to noiseless channels. This theorem is really about repackaging (encoding) data before it is transmitted, so that, when it is transmitted, every datum conveys as much information as possible. We consider the source coding theorem using binary digits below, but the logic of the argument applies equally well to any channel inputs.

Given that a binary digit can convey a maximum of one bit of information, a noiseless channel which communicates $R$ binary digits per second can communicate information at the rate of up to $R$ bits/s. Because the capacity $C$ is the maximum rate at which at which information can be transmitted from input to output, it follows that the capacity of a noiseless channel is numerically equal to the number $R$ of binary digits communicated per second. However, if each binary digit carries less than one bit (e.g. if consecutive output values are correlated) then the rate is less than the channel capacity.

Consider a source which generates a stream of data in the form of signal values $s_1, s_2, \ldots$, with an entropy of $H(S)$ bits per value, and a channel which transmits the corresponding encoded inputs $x_1, x_2, \ldots$, where each input consists of $C$ binary digits. *Shannon's source coding theorem* guarantees that if any source signal $s$ has an entropy $H(s)$ then, when averaged over all values in $s$, (1) just over $H(s)$ binary digits are required to encode each value of $s$, and (2) each value of $s$ cannot be encoded using any fewer than $H(s)$ binary digits. Recalling

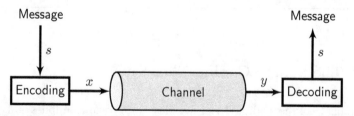

Figure 5.1: The noiseless communication channel. A message (data) is encoded before being used as input to a communication channel. The channel output is decoded by a receiver to recover the message.

the example of the sum of two dice from Section 2.7, a naive encoding would require 3.46 (log 11) binary digits to represent the sum of each throw. However, Shannon's source coding theorem guarantees that an encoding exists such that an average of (just over) 3.27 (i.e. log 9.65) binary digits per value of $s$ will suffice (the phrase 'just over' is an informal interpretation of Shannon's more precise phrase 'arbitrarily close to').

This encoding process yields inputs with a specific distribution $p(X)$, where there are implicit constraints on the form of $p(X)$ (e.g. power constraints). The shape of the distribution $p(X)$ places an upper limit on the entropy $H(X)$ and therefore on the maximum information that each input can carry. Thus, the capacity of a noiseless channel is defined in terms of the particular distribution $p(X)$ which maximises the amount of information per input:

$$C \;=\; \max_{p(X)} H(X) \text{ bits per input.} \qquad (5.1)$$

This states that channel capacity $C$ is achieved by the distribution $p(X)$ which makes $H(X)$ as large as possible (see Section 4.1).

---

**Key point.** Given a continuous variable with a fixed range, the distribution with maximum entropy is the uniform distribution.

---

## 5.2. Why the Theorem is True

In essence, Shannon's source coding theorem is based on the observation that most events that could occur almost certainly do not, and those that do, occur with about the same probability as each other.

47

Consider a source which generates messages, where each message consists of $n$ binary digits $\mathbf{s} = (s_1, \ldots, s_n)$. In principle, the number of different *possible* messages of length $n$ that could be generated by this source is huge, specifically, $m_{max} = 2^n$. But, as we shall see, the number $m$ of different messages actually generated will almost certainly be much, much smaller than $m_{max}$ (i.e. $m \ll m_{max}$). This is the key to Shannon's source coding theorem.

If $n$ is large (i.e. messages are long) then all of the roughly $m$ different messages $\mathbf{s}_1, \ldots, \mathbf{s}_m$ generated by the source will contain about $nP$ binary digits equal to 1. For example, if $P = 1/8 = 0.125$ and $n = 8{,}000$ then the most common messages generated will contain about 1,000 1s. More importantly, as the messages are allowed to get longer, the *law of large numbers* guarantees that almost all messages generated will contain $nP$ 1s.

In fact, Shannon's proof relies on the assumption that messages are very long, but for the purposes of illustration we use a short message.

| Message | | | | | | | | | Codeword | | | |
|---|---|---|---|---|---|---|---|---|---|---|---|---|
| $\mathbf{s}_1$ | 1 | 0 | 0 | 0 | 0 | 0 | 0 | 0 | $\mathbf{x}_1$ | 0 | 0 | 0 |
| $\mathbf{s}_2$ | 0 | 1 | 0 | 0 | 0 | 0 | 0 | 0 | $\mathbf{x}_2$ | 0 | 0 | 1 |
| $\mathbf{s}_3$ | 0 | 0 | 1 | 0 | 0 | 0 | 0 | 0 | $\mathbf{x}_3$ | 0 | 1 | 0 |
| $\mathbf{s}_4$ | 0 | 0 | 0 | 1 | 0 | 0 | 0 | 0 | $\mathbf{x}_4$ | 0 | 1 | 1 |
| $\mathbf{s}_5$ | 0 | 0 | 0 | 0 | 1 | 0 | 0 | 0 | $\mathbf{x}_5$ | 1 | 0 | 0 |
| $\mathbf{s}_6$ | 0 | 0 | 0 | 0 | 0 | 1 | 0 | 0 | $\mathbf{x}_6$ | 1 | 0 | 1 |
| $\mathbf{s}_7$ | 0 | 0 | 0 | 0 | 0 | 0 | 1 | 0 | $\mathbf{x}_7$ | 1 | 1 | 0 |
| $\mathbf{s}_8$ | 0 | 0 | 0 | 0 | 0 | 0 | 0 | 1 | $\mathbf{x}_8$ | 1 | 1 | 1 |
| ... | | | | ... | | | | | Not needed | | | |
| $\mathbf{s}_{256}$ | 1 | 1 | 1 | 1 | 1 | 1 | 1 | 1 | Not needed | | | |

Table 5.1: Why Shannon's source coding theorem is true. Each message (row) from a source contains $n = 8$ binary digits, so up to 256 different messages can be generated. If the probability that each binary digit equals 1 is $P = 1/8$ then most messages contain one 1, so there are effectively only eight different messages, $\mathbf{s}_1, \ldots, \mathbf{s}_8$, which can be represented by eight codewords $\mathbf{x}_1, \ldots, \mathbf{x}_8$, each of which contains three binary digits.

What is plausibly true for large values of $n$ is less plausibly true for small $n$, but the spirit of the Shannon's proof applies to both cases.

If $n=8$ then each message looks like $\mathbf{s} = (s_1, s_2, s_3, s_4, s_5, s_6, s_7, s_8)$, with examples shown in Table 5.1. In principle, the number of different possible messages is $m_{max} = 2^8 = 256$. However, if $P = 0.125$ then $nP = 1$, so the most common messages generated contain one 1, and there are exactly $m = 8$ such messages. Because the probability that each digit in a message is a 1 is independent from the probability that any other digit equals 1 in that message, the probability that a message contains one 1 and seven 0s is not affected by the order in which they occur. Therefore, the probability that a message contains one 1 and seven 0s is the same for all 8 messages. This probability is about 0.4, but we do not care about its value, but we do care that the value is the same for all 8 messages because this means that they all have the same relative frequency in the distribution of all 256 possible messages.

In order to clinch the source coding argument, we need to review a crucial observation. We know that if we have $m = 8$ equally probable outcome values, such as the integers $1, \ldots, 8$, then we can encode each message as one of eight binary codewords $\mathbf{x}_1, \ldots, \mathbf{x}_8$. And because we only need eight such codewords, this means we only need $n_w = \log 8 = 3$ binary digits per codeword, i.e. $\mathbf{x} = (x_1, x_2, x_3)$. In other words, we can send $m = 8$ equally probable integers using exactly $n_w = 3$ binary digits per codeword. Of course, each outcome value does not have to be an integer; it could just as easily be represented as a message of eight binary digits, as shown in Table 5.1. This means that we can encode each of the messages $\mathbf{s}_1, \ldots, \mathbf{s}_8$ with a codeword of three binary digits.

Because the $m = 8$ messages are generated with equal probability, and because we assume that they comprise all of the messages that get generated in practice, it follows that the probability (measured as relative frequency) of each message must be $p = 1/m = 0.125$. If we have $m = 8$ messages, each of which is generated with the same probability $p$, then we know (from Equation 2.13) that the entropy is

$$H(X) \quad = \quad \text{E}[\log(1/p)] = \log m = 3 \text{ bits per message.} \quad (5.2)$$

Thus, even though the source could generate 256 different messages, the 8-binary-digit messages generated have, in practice, an entropy of only $H(X) = 3$ bits per message, and can therefore be encoded with $n_w = H(X) = 3$ binary digits per codeword.

> **Key point.** *Shannon's source coding theorem* states that the number $m$ of different messages actually generated by a source will almost certainly be much smaller than number $m_{max}$ of messages that it could generate (i.e. $m << m_{max}$), and that the messages generated are equiprobable. Consequently, almost all of the messages generated, in practice, from a source can be transmitted using only $H(X) = \log m$ binary digits.

# Chapter 6

# Mutual Information and Noise

The mutual information between two variables, such as a channel input and output, is the average amount of information that each value of the output provides about the input. Somewhat counter-intuitively, this is the same as the average amount of information that each value of the input provides about the output.

Mutual information can be understood intuitively as follows. Before we receive an output $y$, our uncertainty about the input $x$ can be summarised as the input entropy $H(X)$. After we receive an output $y$, our uncertainty should be reduced to some amount, the conditional entropy $H(x|y)$, which is to be read as "the uncertainty (entropy) in $x$ given an output value $y$". So, after receiving an output $y$, our uncertainty about the value of the input $x$ is reduced from $H(x)$ to $H(x|y)$.

However, a reduction in uncertainty amounts to an increase in certainty, or, equivalently, an increase in information. The exact amount of information gained is the difference between the (usually large) uncertainty $H(x)$ we had before we received an output and the (usually smaller) uncertainty $H(x|y)$ we have after receiving an output.

This difference is the mutual information between the input and output,

$$I(X, Y) \quad = \quad H(X) - H(X|Y) \text{ bits.} \tag{6.1}$$

Thus, mutual information is the uncertainty we had before an output was received, minus the the uncertainty we have after an output is received, which corresponds to gaining $I(X, Y)$ bits of information.

If we simply swap the roles of input $X$ and output $Y$ in the preceding section then it follows that

$$I(Y, X) \quad = \quad H(Y) - H(Y|X) \text{ bits,} \tag{6.2}$$

which demonstrates the symmetric nature of mutual information.

> **Key point.** The mutual information between two variables $X$ and $Y$ is the average amount of information that each value of $Y$ provides about $X$. It is also the average amount of information that each value of $X$ provides about $Y$.

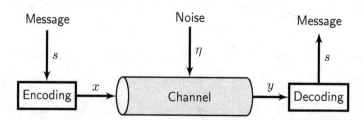

Figure 6.1: The noisy communication channel. A message (data) is encoded before being used as input to a communication channel, which adds noise. The channel output is decoded by a receiver to recover the message.

## 6.1. Mutual Information: Discrete

Consider a channel for which the input is decided by rolling a die with $m_x$ sides. Because the number of equiprobable input states is $m_x$, the input entropy is

$$H(X) \quad = \quad \log m_x \text{ bits.} \tag{6.3}$$

For example, suppose there are $m_x = 8$ equiprobable states, $x_1 = 1$, $x_2 = 2$ and $x_3 = 3$, and so on, so the input entropy is $H(X) = \log 8 = 3$ bits. And if there are $m_\eta = 2$ equiprobable values for the channel noise, say, $\eta_1 = 10$ and $\eta_2 = 20$, then the noise entropy is $H(\eta) = \log 2 = 1.00$ bit.

Now, if the input is $x_1 = 1$ then the output can be one of two equiprobable values, $y_1 = 1 + 10 = 11$ or $y_2 = 1 + 20 = 21$. And if the input is $x_2 = 2$ then the output can be either $y_3 = 12$ or $y_4 = 22$, and so on. Thus, given eight equiprobable input values and two equiprobable

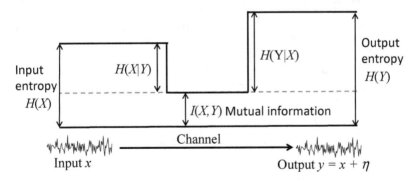

Figure 6.2: Relationships between information theoretic quantities. Noise $\eta$ in the output has entropy $H(\eta) = H(Y|X)$, which represents uncertainty in the output given the input. Noise in the input has entropy $H(X|Y)$, which represents uncertainty in the input given the output. The mutual information is $I(X,Y) = H(X) - H(X|Y) = H(Y) - H(Y|X)$ bits.

noise values, there are $m_y = 16(= 8 \times 2)$ equiprobable output values. So the output entropy is $H(Y) = \log 16 = 4$ bits. However, some of this entropy is noise, so (and this is a crucial point) not all of the output entropy comprises *information about the input*.

In general, the total number $m_y$ of equiprobable output values is $m_y = m_x \times m_\eta$, from which it follows that the output entropy is

$$H(Y) = \log m_x + \log m_\eta \qquad (6.4)$$

$$= H(X) + H(\eta) \text{ bits.} \qquad (6.5)$$

Because we want to explore channel capacity in terms of channel noise, we will pretend to reverse the direction of data along the channel. Consequently, rather than using an output value to estimate the input, we use an input value to estimate the corresponding output value. Accordingly, before we are told the input value, we know that the output can be one of sixteen values, so our uncertainty about the output value is summarised by its entropy, $H(Y) = 4$ bits.

After we have received an input value (e.g. 10), we know the output must be one of two equiprobable values (i.e. 11 or 21), so

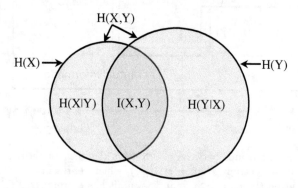

Figure 6.3: Mutual information between related variables $X$ and $Y$. Each circle represents the entropy of one variable.

our uncertainty about the output is reduced from $H(Y)$=4 bits to

$$H(Y|X) \quad = \quad H(\eta) \quad = \quad \log 2 \quad = \quad 1 \text{ bit.} \qquad (6.6)$$

where $H(Y|X)$ is the entropy of the channel noise $H(\eta)$. Equation 6.6 is true for every input, and it is therefore true for the average input. Thus, for each input, we gain an average of

$$H(Y) - H(Y|X) \quad = \quad 4 - 1 \text{ bits} \qquad (6.7)$$

about the output. According to Equation 6.2, this is the amount of mutual information (3 bits) between $X$ and $Y$.

In this example, the noise was designed to allow each output to be uniquely associated with one input value, which is why the mutual information equals the input entropy. But notice that this noise meant that we had to use a channel with an output entropy of 4 bits to transmit 3 bits from input to output; if there had been no noise then we could have used this channel to transmit 4 bits. Thus, noise effectively reduces the amount of information that can be transmitted.

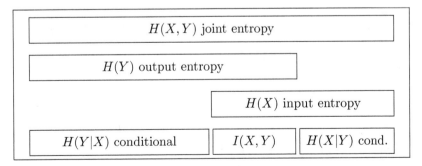

Figure 6.4: The relationship between the input entropy $H(X)$, output entropy $H(Y)$, joint entropy $H(X,Y)$, mutual information $I(X,Y)$, and the conditional entropies $H(X|Y)$ and $H(Y|X)$.

## 6.2. Conditional Entropy

We can prove Equation 6.1 as follows. We begin with the definition of mutual information

$$I(X,Y) \quad = \quad \sum_{i=1}^{m_x}\sum_{j=1}^{m_y} p(x_i, y_j) \log \frac{p(x_i, y_j)}{p(x_i)p(y_j)}. \tag{6.8}$$

$$= \quad E\left[\log \frac{p(x, y)}{p(x)p(y)}\right]. \tag{6.9}$$

According to the *product rule* (see Appendix C), $p(x, y) = p(y|x)p(x)$, where $p(y|x)$ is the *conditional probability* that $Y = y$ given that $X = x$. We can use this to rewrite Equation 6.8 as

$$I(X,Y) \quad = \quad E\left[\log \frac{p(y|x)}{p(y)}\right], \tag{6.10}$$

which in turn can be rewritten as the difference

$$I(X,Y) \quad = \quad E\left[\log \frac{1}{p(y)}\right] - E\left[\log \frac{1}{p(y|x)}\right].$$

The first term on the right is the entropy $H(Y)$ and the final term is the conditional entropy $H(Y|X)$, yielding Equation 6.2, repeated here

$$I(X,Y) \quad = \quad H(Y) - H(Y|X). \tag{6.11}$$

By symmetry, Equation 6.11 implies that

$$I(X,Y) \quad = \quad H(X) - H(X|Y), \tag{6.12}$$

where $H(X|Y)$ is the average uncertainty we have about the value of $X$ after $Y$ is observed. The conditional entropy $H(X|Y)$ is the average uncertainty in $X$ after $Y$ is observed, and is therefore the average uncertainty in $X$ that cannot be attributed to $Y$.

## 6.3. Mutual Information: Continuous

In this section, we explore information in the context of a communication channel which communicates the values of continuous variables. In the context of continuous variables, the mutual information between channel input and output determines the number of different inputs that can be reliably discriminated from a knowledge of the outputs. Specifically, the mutual information is the logarithm of the number $m$ of input values which can be reliably discriminated from a knowledge of the output values ($I = \log m$), where this number is limited by the noise in the channel.

Because mutual information is symmetric (i.e. $I(X,Y) = I(Y,X)$), $m$ is also the logarithm of the number of output values which can be

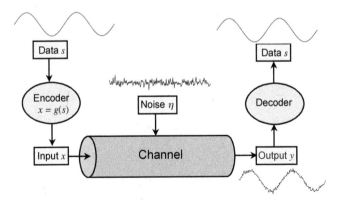

Figure 6.5: The noisy continuous channel. A signal **s** is transformed by an encoding function $\mathbf{x} = g(\mathbf{s})$ before being transmitted through the channel, which corrupts **x** by adding noise $\eta$ to produce the output $\mathbf{y} = \mathbf{x} + \eta$.

reliably discriminated from a knowledge of the input values. Shannon proved that the number $m$ has an upper bound $m_{max}$, for which $C = \log m_{max}$ is the channel capacity.

> **Key point**. Mutual information is the logarithm of the number of input values which can be reliably discriminated from a knowledge of the output values.

**No Infinity**

Notice that both $H(X)$ and $H(X|Y)$ in Equation 6.1 include the infinite constant first encountered in Equation 3.8. However, because this constant has the same value in $H(X)$ and $H(X|Y)$, it cancels out when we subtract one from the other. Thus, the mutual information for continuous variables is not infinite.

**Mutual Information, Conditional Entropy, and Joint Entropy**

So far, we have established three expressions for mutual information,

$$
\begin{align}
I(X,Y) &= H(X) - H(X|Y) \text{ bits} & (6.13) \\
&= H(Y) - H(Y|X) \text{ bits} & (6.14) \\
&= H(X) + H(Y) - H(X,Y) \text{ bits}. & (6.15)
\end{align}
$$

From these, a little algebra yields a fourth expression

$$
I(X,Y) = H(X,Y) - [H(X|Y) + H(Y|X)]. \qquad (6.16)
$$

Thus, the mutual information is that portion of the joint entropy $H(X,Y)$ which is left over once we have removed $[H(X|Y)+H(Y|X)]$, which is the entropy $H(X|Y)$ due to noise in $X$ plus the entropy $H(Y|X)$ due to noise in $Y$.

If we rearrange Equation 6.16 then we obtain

$$H(X,Y) \quad = \quad I(X,Y) + H(X|Y) + H(Y|X). \qquad (6.17)$$

In other words, the joint entropy $H(X,Y)$ acts as an 'entropy container' which consists of three disjoint (i.e. non-overlapping) subsets, as shown in Figures 6.3 and 6.4:

1. the conditional entropy $H(X|Y)$ due to noise in $X$, which is the entropy in $X$ which is not determined by $Y$;

2. the conditional entropy $H(Y|X)$ due to noise in $Y$, which is the entropy in $Y$ which is not determined by $X$;

3. the mutual information $I(X,Y)$, which is the entropy 'shared' by $X$ and $Y$, and which results from the co-dependence of $X$ and $Y$.

> **Key point.** The mutual information is that part of the joint entropy $H(X,Y)$ that is left over once we have removed the part $[H(X|Y) + H(Y|X)]$ due to noise.

## 6.4. Mutual Information Cannot Be Negative

On average, observing an output does not increase uncertainty about the input (even though certain outputs may increase uncertainty). Indeed, it can be shown (see Reza (1961)[20]) that the entropy of $X$ given $Y$ cannot be greater than the entropy of $X$,

$$H(X|Y) \quad \leq \quad H(X) \text{ bits}, \qquad (6.18)$$

with equality only if $X$ and $Y$ are independent. From Equation 6.11, it follows that mutual information is positive, unless $X$ and $Y$ are independent, in which case it is zero.

# Chapter 7

# Noisy Channels

## 7.1. The Noisy Channel Coding Theorem

Remarkable as it is, Shannon's source coding theorem ignores the effects of noise. In contrast, an informal summary of Shannon's (equally remarkable) *noisy channel coding theorem* states that: *it is possible to use a noisy channel to communicate information almost without error at a rate arbitrarily close to the channel capacity of $C$ bits/s, but it is not possible to communicate information at a rate greater than $C$ bits/s.*

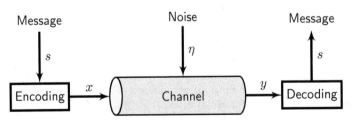

Figure 7.1: The noisy communication channel. A message (data) is encoded before being used as input to a communication channel, which adds noise. The channel output is decoded by a receiver to recover the message.

The capacity of a noisy channel is defined as

$$C = \max_{p(x)} I(X,Y) \qquad (7.1)$$

$$= \max_{p(x)} [H(Y) - H(Y|X)] \text{ bits.} \qquad (7.2)$$

If there is no noise (i.e. if $H(Y|X) = 0$) then this reduces to Equation 5.1, which is the capacity of a noiseless channel. The *data processing inequality* states that, no matter how sophisticated any device is, the amount of information $I(X,Y)$ in the output about the input cannot be greater than the amount of information $H(X)$ in the input.

## 7.2. Why the Theorem is True

Describing Shannon's proof in detail would require more mathematical tools than we have available here, so this is a brief summary which gives a flavour of his proof.

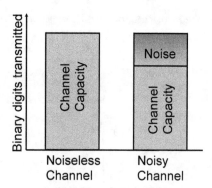

Figure 7.2: The *channel capacity* of noiseless and noisy channels is the maximum rate at which information can be communicated. If a noiseless channel communicates data at 10 binary digits/s then its capacity is $C = 10$ bits/s. The capacity of a noiseless channel is numerically equal to the rate at which it communicates binary digits, whereas the capacity of a noisy channel is less than this because it is limited by the amount of noise in the channel.

Consider a discrete or continuous channel with a fixed amount of channel noise and capacity $C$. We have a set of $N$ messages $\mathbf{s}_1, \ldots, \mathbf{s}_N$ which have been encoded to produce inputs $\mathbf{x}_1, \ldots, \mathbf{x}_N$ such that the entropy $H$ of these inputs is less than the channel capacity $C$. Now imagine that we construct a bizarre codebook in which each randomly chosen input $\mathbf{x}_i$ gets interpreted as a fixed, but randomly chosen, output $\mathbf{y}_i$. By chance, some outputs will get assigned the same, or very similar, inputs, and *vice versa*, leading to a degree of cross-talk. Consequently, when we use this codebook to decode outputs, we are bound to misclassify a proportion of them. This proportion is the error rate of the codebook. We then repeat this madness until we have recorded the error rate of all possible codebooks.

Shannon proved that, provided $H \leq C$, when averaged over all possible codebooks, the average error rate approaches zero as the length of the inputs $\mathbf{x}$ increases. Consequently, if we make use of long inputs, so that the *average* error rate $\epsilon$ is small, then *there must exist at least one codebook which produces an error as small as $\epsilon$*. Notice that if all codebooks produce the same error rate $\epsilon$ then the average error rate is also $\epsilon$, but if just one codebook has an error greater than $\epsilon$ then at least one codebook has an error rate smaller than $\epsilon$.

As Pierce (1961)[19] notes, some people regard the logic which underpins Shannon's proof as weird, but such an outrageous proof also gives some insight into the distinctive mind which created it.

## 7.3. The Gaussian Channel

If the noise values in a channel are drawn independently from a Gaussian distribution (i.e. $\eta \sim \mathcal{N}(\mu_\eta, v_\eta)$, as defined in Equation 4.7) then this defines a *Gaussian channel*.

Given that $Y = X + \eta$, if we want $p(Y)$ to be Gaussian then we should ensure that $p(X)$ and $p(\eta)$ are Gaussian, because the sum of two independent Gaussian variables is also Gaussian[20]. So $p(X)$ must be (iid) Gaussian in order to maximise $H(X)$, which maximises $H(Y)$, which maximises $I(X,Y)$. Thus, if each input, output, and noise variable is (iid) Gaussian then the average amount of information communicated per output value is the channel capacity, so that $I(X,Y) = C$ bits. This is informal statement of *Shannon's continuous noisy channel coding theorem for Gaussian channels* can be used to express capacity in terms of the input, output, and noise.

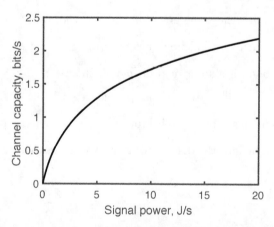

Figure 7.3: Increasing signal power $S$ (Joules per second) has diminishing returns on the capacity of Gaussian channel capacity (bits per second) (Equation 7.7, with noise power (variance) $N = 1$).

If the channel input $x \sim \mathcal{N}(\mu_x, v_x)$ then the continuous analogue (integral) of Equation 2.13 yields the *differential entropy*

$$H(X) = (1/2) \log 2\pi e v_x \text{ bits.} \qquad (7.3)$$

The distinction between differential entropy and entropy effectively disappears when considering the difference between entropies, and we will therefore find that we can safely ignore this distinction here. Given that the channel noise is iid Gaussian, its entropy $H(\eta)$ is

$$H(Y|X) = (1/2) \log 2\pi e v_\eta \text{ bits.} \qquad (7.4)$$

Because the output is the sum $Y = X + \eta$, it is also iid Gaussian with variance $v_y = v_x + v_\eta$, and its entropy is

$$H(Y) = (1/2) \log 2\pi e (v_x + v_\eta) \text{ bits.} \qquad (7.5)$$

Substituting Equations 7.4 and 7.5 into Equation 6.11 yields

$$I(X, Y) = \frac{1}{2} \log \left( 1 + \frac{v_x}{v_\eta} \right) \text{ bits,} \qquad (7.6)$$

which allows us to choose one out of $m = 2^I$ equiprobable values. For a Gaussian channel, $I(X, Y)$ attains its maximal value of $C$ bits.

The variance of any signal with a mean of zero is equal to its *power*, which is the rate at which energy is expended per second, and the physical unit of power is measured in *Joules* per second (J/s) or *Watts*, where 1 Watt $= 1$ J/s. Accordingly, the signal power is $P = v_x$ J/s, and the noise power is $N = v_\eta$ J/s. This yields Shannon's famous equation for the capacity of a Gaussian channel:

$$C \;=\; \frac{1}{2}\log\left(1 + \frac{P}{N}\right) \text{ bits,} \qquad (7.7)$$

where the ratio of variances $P/N$ is the *signal-to-noise ratio* (SNR), as in Figure 7.3.

> **Key point.** Channel noise effectively discretises a continuous Gaussian output distribution into a discrete distribution.

Given that we want to transmit as much information as possible for each Watt of power expended, should we increase the number $n$ of values in each input vector **x**, or should we increase the signal power $P$ by increasing the amplitude of each value in **x**?

Clearly, doubling $n$ doubles the length of **x**, which doubles the power required to transmit **x**, but it also doubles the amount of information transmitted. In contrast, doubling the signal power $P$ increases, but does not double, the amount of information transmitted, as can be seen from Figure 7.3. Thus, given a choice between increasing the number $n$ of values transmitted and increasing the amplitude of transmitted values (i.e. signal power $P$), we should increase $n$. Although we have not covered the topic of *signal bandwidth* in this book, the above result implies that if we have a choice between boosting signal power and increasing the bandwidth then we should increase the bandwidth.

# Chapter 8

# Rate Distortion Theory

Nature dictates that a signal can either be represented expensively with high fidelity, or cheaply with low fidelity, but it cannot be represented cheaply with high fidelity. Rate distortion theory defines the lowest price that must be paid at every possible value of fidelity. And the only currency Nature accepts as payment is information.

Given a noiseless channel with a fixed capacity $C$, we can keep increasing the amount of data transmitted, and we can expect perfect fidelity at the output until the entropy of the input approaches the channel capacity. Up to this point Shannon's source coding theorem guarantees that the uncertainty about the input given the output, or equivocation, can be zero. However, if we attempt to transmit more data through the channel then the equivocation must increase. Crucially, the increase in equivocation exactly matches the extent to which the input entropy exceeds the channel capacity. Consequently, the transmission rate $R$ remains approximately equal to the channel capacity $C$, irrespective of how much the input entropy exceeds the channel capacity, as shown in Figure 8.1.

Consider a channel with capacity $C$, for which the input is one element of an $n$-element vector of iid variables,

$$\mathbf{X} \; = \; (X_1, \ldots, X_n), \tag{8.1}$$

which has entropy $H(\mathbf{X}) = nH(X)$. In the simplest case, the channel output is also one element of an $n$-element vector

$$\hat{\mathbf{X}} \; = \; (\hat{X}_1, \ldots, \hat{X}_n). \tag{8.2}$$

In order to transmit the vector $\mathbf{X}$ through a channel without error, we must transmit at the *rate* $R = H(X)$ bits/element, so the channel capacity must be at least $C = R$ bits/element.

However, if we are willing to tolerate some error or *distortion* $D$ in the output $\hat{\mathbf{X}}$ then the rate, and therefore the channel capacity $C$,

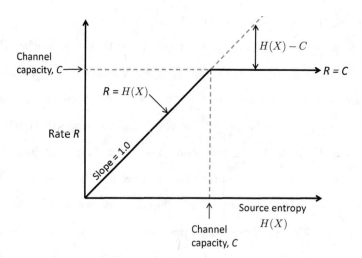

Figure 8.1: Graphical representation of rate distortion theory for a channel with input $X$, output $Y$, and capacity $C$. If $H(X) < C$ then the transmission rate $R \approx H(X)$. If $H(X) > C$ then $R \approx C$, and the dashed diagonal line represents the extent to which the input entropy $H(X)$ exceeds $C$, which results in an equivocation $H(X|Y) \geq H(X) - C$.

can be less than $R=H(X)$. As the amount of distortion is increased, the uncertainty of the input $X$ given an output $\hat{X}$ also increases, and a convenient measure of this uncertainty is the *equivocation* $E=H(X|\hat{X})$.

Rate distortion theory is, to some extent, an extension of the noisy channel coding theorem, which places a lower bound on equivocation,

$$H(X|\hat{X}) \quad \geq \quad H(X) - C \text{ bits,} \qquad (8.3)$$

where the channel capacity $C$ is equal to a particular rate $R$. Shannon[25] succinctly summarised the implication of this bound as follows,

> If an attempt is made to transmit at a higher rate than $C$, say $C + E$, then there will necessarily be an equivocation equal to or greater than the excess $E$. Nature takes payment by requiring just that much uncertainty.

**Notation.** For the sake of brevity, we avoid the verbal sumersaults required to express the limiting behaviour of information rates in this chapter. Accordingly, we will sometimes take the liberty of replacing phrases such as "$R \rightarrow C$ as $n \rightarrow \infty$" with "$R = C$".

> **Key point.** Rate distortion theory specifies the smallest channel capacity or rate $R$ required to recover the channel input with a level of distortion no larger than $D$.

### The Rate Distortion Function

Rate distortion theory defines the *rate distortion function* $R(D)$ as the smallest number of bits or rate $R$ that must be allocated to represent $X$ as a distorted version $\hat{X}$ for a given level of distortion $D$. As an example, distortion can be measured as the mean squared difference

$$D \;=\; \mathrm{E}[(X - \hat{X})^2], \qquad (8.4)$$

between the input $X$ and the output $\hat{X}$, for which a rate distortion function is shown in Figure 8.3.

Rate distortion theory also defines the *distortion rate function $D(R)$* as the smallest distortion $D$ that is consistent with a given rate $R$.

> **Key point.** We can have low distortion if we use a large number of bits or *rate* to represent data. But if we use a small number of bits then we must have high distortion. Thus, we can have low distortion at a high rate, or high distortion at a low rate, but we cannot have low distortion at a low rate.

## 8.1. The Binary Rate Distortion Function

Consider a binary vector $\mathbf{x}$, where the probability of each element being a 1 is $p$. If the distortion measure is

$$d(x, \hat{x}) \;=\; \begin{cases} 0, & \text{if } x = \hat{x} \\[2mm] 1, & \text{if } x \neq \hat{x}, \end{cases} \qquad (8.5)$$

and the mean distortion is $D=\mathrm{E}[d(X,\hat{X})]$. It can be shown that the rate distortion function is

$$R(D) = \begin{cases} H(p) - H(D) \text{ bits}, & \text{if } 0 \leq D \leq \min\{p, 1-p\} \\[2mm] 0 \text{ bits}, & \text{if } D > \min\{p, 1-p\}, \end{cases} \qquad (8.6)$$

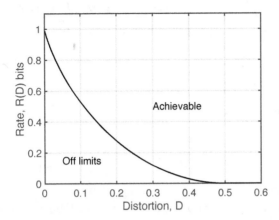

Figure 8.2: Rate distortion function for a binary source with $p = 0.5$ (Equation 8.6). Only the region above the curve corresponds to $(R, D)$ values that can be achieved.

as shown in Figure 8.2 for $p = 0.5$. Notice that the rate is $R(D) = 1$ bit when $D = 0$, as we should expect from a noiseless binary variable.

## 8.2. The Gaussian Rate Distortion Function

Consider a vector $\mathbf{x}$ of $n$ elements $x$ in which each element is drawn from a Gaussian distribution $x \sim \mathcal{N}(0, \sigma^2)$. If the distortion is $d(x, \hat{x}) = (x - \hat{x})^2$ then

$$d(\mathbf{X}, \hat{\mathbf{X}}) = \frac{1}{n} \sum_{i=1}^{n} (x_i - \hat{x}_i)^2, \tag{8.7}$$

so the mean distortion is $D = \mathrm{E}[d(\mathbf{X}, \hat{\mathbf{X}})]$. It can be shown that if a source has variance $\sigma^2$ then the rate distortion function is

$$R(D) = \begin{cases} \frac{1}{2} \log \frac{\sigma^2}{D}, & \text{if } 0 \le D \le \sigma^2 \\ 0, & \text{if } D > \sigma^2. \end{cases} \tag{8.8}$$

For example, if $\sigma^2 = 1$ then $R(D) = 1/2 \log 1/D$ bits, as shown in Figure 8.3 for $\sigma^2 = 1$. Notice that $R(D) \to \infty$ bits as $D \to 0$, as we should expect given that the exact value of a real variable conveys an infinite amount of information (see Section 3.1).

Re-arranging Equation 8.8 yields the distortion rate function

$$D(R) \quad = \quad \sigma^2/2^{2R}. \tag{8.9}$$

This implies that each extra bit added to the rate $R$ decreases the mean distortion by a factor of $2^2{=}4$. For example, if the encoding function yields $\hat{\mathbf{x}}$ such that each element of $\hat{\mathbf{x}}$ has entropy $R{=}1$ bit then the smallest mean squared error is $D(R){=}0.25\sigma^2$ (see Section ??).

As noted above, rate distortion theory has many applications in telecommunications. But it has also been applied to other systems, such as evolution, human perception, and navigation in bacteria. For reviews of information theory and rate distortion theory in biology, see Taylor *et al* (2007)[28], Tkačik and Bialek (2012)[4], and Bialek (2016)[29].

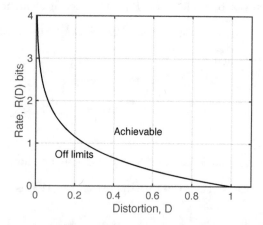

Figure 8.3: Rate distortion function for a Gaussian source with variance $\sigma^2 = 1$ (Equation 8.8). Only the region above the curve corresponds to $(R, D)$ values that can be achieved.

Whereas information theory assumes an ideal world in which channel capacity is sufficient to transmit messages without error (in principle), rate distortion theory relies on the more realistic assumption that channel capacity is never sufficient, and therefore some degree of error must be tolerated. Even though rate distortion theory does not provide a 'free lunch', in terms of information and distortion, it does specify the lowest possible information price that must be paid for a given level of distortion.

## 8.3. Conclusion

In 1986, the physicist John Wheeler said:

> It is my opinion that everything must be based on a simple idea. And ... this idea, once we have finally discovered it, will be so compelling, so beautiful, that we will say to one another, yes, how could it have been any different?

So compelling, and so beautiful: information theory represents a fundamental insight that must surely rank as a candidate for Wheeler's "simple idea". Indeed, after many years of studying physics and information theory, Wheeler came up with a proposal which is both radical and intriguing:

> ... the universe is made of information; matter and energy are only incidental.

Insofar as it must be made of something, a universe in which all forms of energy and matter are simply different manifestations of pure information might be as sublime as this one.

# Further Reading

Applebaum D (2008)[1]. Probability and Information: An Integrated Approach. *A thorough introduction to information theory, which strikes a good balance between intuitive and technical explanations.*

Avery J (2012)[2]. Information Theory and Evolution. *An engaging account of how information theory is relevant to a wide range of natural and man-made systems, including evolution, physics, culture and genetics. Includes interesting background stories on the development of ideas within these different disciplines.*

Baeyer HV (2005)[3]. Information: The New Language of Science *Erudite, wide-ranging, and insightful account of information theory. Contains no equations, which makes it very readable.*

Cover T and Thomas J (1991)[6]. Elements of Information Theory. *Comprehensive, and highly technical, with historical notes and an equation summary at the end of each chapter.*

Ghahramani Z (2002). Information Theory. Encyclopedia of Cognitive Science. *An excellent, brief overview of information.*

Gibson JD (2014)[10]. Information Theory and Rate Distortion Theory for Communications and Compression *A short (115 pages) introduction to information theory and rate distortion theory that is both formal and reasonably accessible, with an emphasis on key theorems and proofs.*

Gleick J (2012)[11]. The Information. *An informal introduction to the history of ideas and people associated with information theory.*

Guizzo EM (2003)[12]. The Essential Message: Claude Shannon and the Making of Information Theory. Master's Thesis, Massachusetts Institute of Technology. *One of the few accounts of Shannon's role in the development of information theory. See* http://dspace.mit.edu/bitstream/handle/1721.1/39429/54526133.pdf.

MacKay DJC (2003)[16]. Information Theory, Inference, and Learning Algorithms. *The modern classic on information theory. A very readable text that roams far and wide over many topics.*

*The book's web site (below) also has a link to an excellent series of video lectures by MacKay. Available free online at* http://www.inference.phy.cam.ac.uk/mackay/itila/.

Pierce JR (1980)[19]. An Introduction to Information Theory: Symbols, Signals and Noise. Second Edition. *Pierce writes with an informal, tutorial style of writing, but does not flinch from presenting the fundamental theorems of information theory. This book provides a good balance between words and equations.*

Reza FM (1961)[20]. An Introduction to Information Theory. *A comprehensive and mathematically rigorous book; it should be read only after first reading Pierce's more informal text.*

Shannon CE and Weaver W (1949)[25]. The Mathematical Theory of Communication. University of Illinois Press. *A surprisingly accessible book, written in an era when information theory was known only to a privileged few.* This book can be downloaded from http://cm.bell-labs.com/cm/ms/what/shannonday/paper.html

Shannon CE (1959)[24]. Coding theorems for a discrete source with a fidelity criterion. *Shannon's 22 page paper on rate distortion theory.*

Stone JV (2022)[27]. Information Theory: A Tutorial Introduction. *Information theory explained at a more leisurely pace than the current text.*

For the complete novice, the videos at the online Kahn Academy provide an excellent introduction. Additionally, the Scholarpedia web page by Latham and Rudi provides a lucid technical account of mutual information: http://www.scholarpedia.org/article/Mutual_information.

The Bit Player is a fine documentary of Shannon's life and works (available at cinema special showings).

Fritterin' Away Genius. Tim Harford's podcast provides insight into Shannon's playful personality. https://omny.fm/shows/cautionary-tales-with-tim-harford/fritterin-away-genius

Historical perspective provided by an interview with Shannon in 1982: https://www.ethw.org/Oral-History:Claude_E._Shannon.

# Appendix A

# Glossary

**alphabet** A set of symbols used to construct a message. For example, the numbers 1 to 6 define a discrete variable which has an alphabet of possible values $A_s = \{1, 2, 3, 4, 5, 6\}$, and a message could then be either a single symbol, such as $s = 4$, or an ordered list of symbols, such as $\mathbf{s} = (1, 1, 6, 3, 2, 3, 1, 1, 4)$.

**average** Given a variable $x$, the average, mean or expected value of $n$ values is

$$\overline{x} = \frac{1}{n} \sum_{j=1}^{n} x_j. \qquad (A.1)$$

**binary digit** A binary digit can be either a 0 or a 1.

**binary number** A number that consists of binary digits (e.g. 1001).

**bit** A fundamental unit of information, often confused with a binary digit (see Section 2.3). A bit provides enough information for one of two equally probable alternatives to be specified.

**byte** An ordered set of 8 binary digits.

**capacity** The capacity of a communication channel is the maximum rate at which it can communicate information from its input to its output. Capacity can be specified either in terms of information communicated per second (e.g. bits/s), or in terms of information communicated per symbol (e.g. bits/symbol).

**channel** A conduit for communicating data from its input to its output.

**code** A code consists of a set of symbols or messages, an encoder (which maps symbols to channel inputs), a decoder (which maps channel outputs to inputs).

**codebook** The set of codewords produced by a given encoder.

**codeword** Each symbol $s$ in a message is encoded before transmission as a codeword $x$.

**conditional probability** The probability that the value of one random variable $Y$ has the value $y$ given that the value of another random variable $X$ has the value $x$, written as $p(Y = y | X = x)$ or $p(y|x)$.

**conditional entropy** Given two random variables $X$ and $Y$, the average uncertainty regarding the value of $Y$ when the value of $X$ is known, $H(Y|X) = \mathrm{E}[\log(1/p(y|x))]$ bits.

**continuous** Whereas a discrete variable adopts a discrete number of values, a continuous variable can adopt any value (e.g. a decimal).

**differential entropy** The expected value of a continuous random variable, $\mathrm{E}[\log(1/p(x))]$.

**discrete** Elements of set that are clearly separated from each other, like a list of integers, are called discrete. See also continuous.

**encoding** Before a message is transmitted, it is or encoded as an input sequence. Ideally, the encoding process ensures that each element of the encoded message conveys as much information as possible.

**equivocation** Average uncertainty of the value of the channel input $x$ when the output $y$ is known, measured as conditional entropy $H(X|Y)$.

**entropy** The entropy of a variable is a measure of its overall variability.

**expected value** See average.

**histogram** If we count the number of times the value of a discrete variable adopts each of a number of values then the resultant set of counts defines a histogram. If each count is divided by the total number of counts then the resultant set of proportions defines a *normalised histogram*.

**iid** If values are chosen independently (i.e. 'at random') from a single probability distribution then they are said to be iid (*independent and identically distributed*).

**independence** If two variables $X$ and $Y$ are independent then the value $x$ of $X$ provides no information regarding the value $y$ of the other variable $Y$, and *vice versa*.

**information** The amount of information conveyed by a discrete variable $X$ which has a value $X = x$ is $h(x) = \log(1/p(x))$. The average amount of information conveyed by each value of $X$ is its entropy $H(X) = \sum p(x_i) \log(1/p(x_i))$.

**joint probability** The probability that two or more quantities simultaneously adopt specified values. For example, the probability that one die yields $x_3 = 3$ and another yields $y_4 = 4$ is the joint probability $p(x_3, y_4) = 1/36$.

**law of large numbers** Given a variable $X$ with a mean $\mu$, the mean of a sample of $n$ values converges to $\mu$ as the number of values in that sample approaches infinity; that is, $\mathrm{E}[X] \to \mu$ as $n \to \infty$.

**logarithm** Given a number $x$ which we wish to express as a logarithm with base $a$, $y = \log_a x$ is the power to which we have to raise $a$ in order to get $x$. See Section 2.2.

**mean** See average.

**message** A sequence of symbols or values, represented in bold **s** or non-bold $s$, according to context.

**mutual information** The reduction in uncertainty $I(X,Y)$ regarding the value of one variable $Y$ induced by knowing the value of another variable $X$. Mutual information is symmetric, so $I(X,Y) = I(Y,X)$.

**noise** The random 'jitter' that is part of a measured quantity.

**outcome** In this text, the term outcome refers to a single instance of a physical outcome, like the pair of numbers showing after a pair of dice is thrown. In terms of random variables, an outcome is the result of a single experiment.

**outcome value** In this text, the term outcome value refers to the numerical value assigned to a single physical outcome. For example, if a pair of dice is thrown then the outcome $(x_1, x_2)$ comprises two numbers, and the outcome value can be defined as the sum of these two numbers, $x = x_1 + x_2$. In terms of the random variable $X$, the outcome value is the numerical value assigned to the outcome $(x_1, x_2)$, written as $x = X(x_1, x_2)$.

**precision** An indication of the granularity or resolution with which a variable can be measured, formally defined as the inverse of variance (i.e. precision=1/variance).

**probability** There are many definitions of probability. The two main ones are (using coin bias as an example): (1) Bayesian: an observer's estimate of the probability that a coin will land heads up is based on all the information the observer has, including the proportion of times it was observed to land heads up in the past. (2) Frequentist: the probability that a coin will land heads up is given by the proportion of times it lands heads up, when measured over many flips.

**probability density function (pdf)** The probability density function (pdf) $p(X)$ of a continuous random variable $X$ defines the probability density of each value of $X$. Loosely speaking, the probability that $X=x$ can be considered as the probability density $p(x)$.

**probability function (pf)** A function $p(X)$ of a discrete random variable $X$ defines the probability of each value of $X$. The probability that $X = x$ is $p(X = x)$ or, more succinctly, $p(x)$. This is called a *probability mass function* (pmf) in some texts.

**product rule** ADD THIS.

**random variable (RV)** The concept of a random variable $X$ can be understood from a simple example, such as the throw of a pair of dice.

Each physical outcome is a pair of numbers $(x_a, x_b)$, which is assigned a value (e.g., $x = x_a + x_b$) which is taken to be the value of the random variable, so that $X = x$. The probability of each value is defined by a probability distribution $p(X) = \{p(x_1), p(x_2), \dots\}$.

**rate** In information theory, this is the rate $R$ at which information is transmitted from the input $X$ to the output $Y$ of a channel, $R = H(X) - H(X|Y)$ bits/s, which cannot exceed the channel capacity $C$. In rate distortion theory, $\hat{X}$ is a distorted version $X$, and the rate $R$ is the entropy $H(\hat{X})$ of $\hat{X}$, which is the number of bits allocated to represent element $X$ of a signal $\mathbf{X} = (X_1, \dots, X_n)$.

**rate distortion function** The smallest number of bits or rate $R$ that must be allocated to represent $X$ as as a distorted version $\hat{X}$ for a given level of distortion $D$.

**rate equivocation function** The smallest number of bits or rate $R$ that must be allocated to represent $X$ as a distorted version $\hat{X}$ for a given level of equivocation $E = H(X|\hat{X})$.

**rate distortion theory** Provides a theoretical bound on the information required to represent a signal with a given level of distortion.

**redundancy** Given an ordered set of values of a variable (e.g. in an image or sound), if a value can be obtained from a knowledge of other values then it is redundant.

**relative frequency** Frequency of occurrence, expressed as a proportion. For example, out of every 10,000 English letters, 1,304 of them are the letter E, so the relative frequency of E is 1304/10000=0.134.

**standard deviation** The square root $\sigma$ of the variance of a variable.

**theorem** A mathematical statement which has been proven to be true.

**transmission rate** See rate.

**uncertainty** In this text, uncertainty refers to the surprisal (i.e. $\log(1/p(x))$) of a variable $X$.

**variable** A variable is essentially a 'container', usually for one number.

**variance** The variance is a measure of how 'spread out' the values of a variable are. Given a sample of $n$ values of a variable $x$ with a sample mean $\bar{x}$, the estimated variance $\hat{v}_x$ of $x$ is

$$\hat{v}_x = \frac{1}{n} \sum_{j=1}^{n} (x_j - \bar{x})^2. \tag{A.2}$$

# Appendix B

# Mathematical Symbols

$\hat{}$ the hat symbol is used to indicate an *estimated* value. For example, $\hat{v}_x$ is an estimate of the variance $v_x$.

$|x|$ indicates the absolute value of $x$ (e.g. if $x = -3$ then $|x| = 3$).

$\leq$ if $x \leq y$ then $x$ is less than or equal to $y$.

$\geq$ if $x \geq y$ then $x$ is greater than or equal to $y$.

$\approx$ means 'approximately equal to'.

$\sim$ if a random variable $X$ has a distribution $p(X)$ then this is written as $X \sim p(X)$.

$\infty$ infinity.

$\propto$ indicates *proportional to*.

$\Delta$ Greek upper case letter delta, denotes a small increment.

$\epsilon$ Greek letter epsilon, denotes a small quantity.

$\eta$ Greek letter eta (pronounced eater), denotes a single value of the noise in a measured quantity.

$\eta$ large Greek letter eta, used in this text to denote a random variable for noise.

$\mu$ Greek letter mu (pronounced mew), the mean value of a variable.

$\sigma$ Greek letter sigma, denotes the standard deviation of a distribution.

$\sum$ the capital Greek letter sigma represents summation. For example, if we represent the $n = 3$ numbers 2, 5 and 7 as $x_1 = 2$, $x_2 = 5$, $x_3 = 7$

then their sum $x_{sum}$ is

$$x_{sum} = \sum_{i=1}^{n} x_i = x_1 + x_2 + x_3 = 2 + 5 + 7 = 14.$$

The variable $i$ is counted up from 1 to $n$, and, for each $i$, the term $x_i$ adopts a new value and is added to a running total.

$A$ the set or alphabet of different values of a random variable. For example, if the random variable $X$ can adopt one of $m$ different values then the set $A_x$ is $A_x = \{x_1, \ldots, x_m\}$.

$C$ channel capacity, the maximum rate at which information can be communicated through a given channel, usually measured in bits per second (bits/s).

$D$ mean distortion between elements of the sequences $\mathbf{x}$ and $\hat{\mathbf{x}}$.

$d$ distortion of corresponding elements of the sequences $\mathbf{x}$ and $\hat{\mathbf{x}}$.

$e$ constant, equal to 2.7 1828 1828 .... Base of natural logarithms, so that $\ln e^x = x$.

$E$ the mean, average, or *expected value* of a variable $X$, written as $E[X]$.

$E$ equivocation, or conditional entropy $H(X|Y)$.

$g$ encoding function, which transforms a message of symbols $\mathbf{s} = (s_1, \ldots, s_k)$ into channel inputs $\mathbf{x} = (x_1, \ldots, x_n)$, so $\mathbf{x} = g(\mathbf{s})$.

$h(x)$ Shannon information, uncertainty, or surprisal, $\log(1/p(x))$, of the value $x$.

$H(X)$ entropy of $X$, which is the average Shannon information of the probability distribution $p(X)$ of the random variable $X$.

$H(X|Y)$ conditional entropy of the conditional probability distribution $p(X|Y)$ of values adopted by the variable $X$ given values of the variable $Y$. This is the average uncertainty in the value of $X$ after the value of $Y$ is observed.

$H(Y|X)$ conditional entropy of the conditional probability distribution $p(Y|X)$ of values adopted by the variable $Y$ given values of the variable $X$. This is the average uncertainty in the value of $Y$ after the value of $X$ is observed.

$H(X,Y)$ entropy of the joint probability distribution $p(X,Y)$ of the variables $X$ and $Y$.

$I(X,Y)$ mutual information between $X$ and $Y$, the average number of bits provided by each value of $Y$ about the value of $X$, and *vice versa*.

82

ln $x$ natural logarithm (log to the base $e$) of $x$.

log $x$ logarithm of $x$. Logarithms use base 2 in this text, and base is indicated with a subscript if the base is unclear (e.g. $\log_2 x$). Natural logarithms are logarithms to the base $e$, and are usually written as ln $x$.

$m$ number of different possible messages, input values, codewords, or symbols in an alphabet.

$N$ noise variance in Shannon's fundamental equation for channel capacity $C = \frac{1}{2}\log(1 + P/N)$.

$n$ the number of observations in a data set (e.g. coin flip outcomes), or elements in a message, or codewords in an encoded message.

$p(X)$ the probability distribution of the random variable $X$.

$p(x)$ the probability (density) that the random variable $X = x$.

$p(X, Y)$ the joint probability distribution of the random variables $X$ and $Y$. For discrete variables this is called the *joint probability function* (pf) of $X$ and $Y$, and for continuous variables it is called the *joint probability density function* (pdf) of $X$ and $Y$.

$p(x, y)$ the *joint probability* that the random variables $X$ and $Y$ have the values $x$ and $y$, respectively.

$p(x|y)$ the conditional probability that the random variable $X = x$ given that $Y = y$.

$R$ the rate at which information is communicated, usually measured in bits per second (bits/s), $R = H(X) - H(X|Y)$.

$S$ a random variable. The probability that $S$ adopts a value $s$ is defined by the value of the probability distribution $p(S)$ at $S = s$.

$s$ a value of the random variable $S$, used to represent a message.

$v_x$ if $X$ has mean $\mu$ then the variance of $X$ is $v_x = \sigma_x^2 = E[(\mu - x)^2]$.

$X$ a random variable. The probability that $X$ adopts a specific value $x$ is defined by the value of the probability distribution $p(X)$ at $X = x$.

$X^{\triangle}$ a variable which has been quantised into intervals (e.g. histogram bins) of width $\Delta x$.

$\hat{X}$ random variable used to represent the distorted version of $X$.

$x$ a value of the random variable $X$, used to represent a channel input.

$\mathbf{x}$ a vector or permutation (round brackets $(x_1, \ldots, x_n)$) or combination (curly brackets $\{x_1, \ldots, x_n\}$) of $x$ values.

## Mathematical Symbols

$\hat{\mathbf{x}}$ a sequence $(\hat{x}_1, \ldots, \hat{x}_n)$ of $x$ values.

$\mathbf{X}$ a sequence $(X_1, \ldots, X_n)$ of $n$ random variables.

$\hat{\mathbf{X}}$ a sequence $(\hat{X}_1, \ldots, \hat{X}_n)$ of $n$ random variables.

$Y$ a random variable. The probability that $Y$ adopts a specific value $y$ is defined by the value of the probability distribution $p(Y)$ at $Y = y$.

$y$ a value of the random variable $Y$, used to represent a channel output.

# Appendix C

# The Rules of Probability

## Independent Outcomes

If a set of individual outcomes are independent then the probability of that outcome set is obtained by multiplying the probabilities of the individual outcomes together.

For example, consider a coin for which the probability of a head $x_h$ is $p(x_h) = 0.9$ and the probability of a tail $x_t$ is $p(x_t) = (1-0.9) = 0.1$. If we flip this coin twice then there are four possible pairs of outcomes: two heads $(x_h, x_h)$, two tails $(x_t, x_t)$, a head followed by a tail $(x_h, x_t)$, and a tail followed by a head $(x_t, x_h)$.

The probability that the first outcome is a head and the second outcome is a tail can be represented as a *joint probability* $p(x_h, x_t)$ (More generally, a joint probability can refer to any pair of variables, such as $X$ and $Y$.)

In order to work out some averages, imagine that we perform 100 pairs of coin flips. We label each flip according to whether it came first or second within its pair, so we have 100 *first flip* outcomes, and 100 corresponding *second flip* outcomes (see Table C.1).

| Outcome | $h$ | $t$ | $\{h, h\}$ | $\{t, t\}$ | $(h, t)$ | $(t, h)$ | $\{t, h\}$ |
|---------|------|------|-----------|-----------|----------|----------|-----------|
| $N$ | 90 | 10 | 81 | 1 | 9 | 9 | 18 |
| $N/100$ | 0.90 | 0.10 | 0.81 | 0.01 | 0.09 | 0.09 | 0.18 |

Table C.1: The number $N$ and probability $N/100$ of each possible outcome from 100 pairs of coin flips of a coin which lands heads up 90% of the time. Ordered sequences or permutations are written in round brackets '()', whereas unordered sets or combinations are written in curly brackets '{}'.

85

Given that $p(x_h) = 0.9$, we expect 90 heads and 10 tails within the set of 100 first flips, and the same for the set of 100 second flips. But what about the number of pairs of outcomes?

For each head obtained on the first flip, we can observe the corresponding outcome on the second flip, and then add up the number of pairs of each type (e.g. $x_h, x_h$). We already know that, within the set of 100 first flip outcomes, the average number of heads is

$$90 \quad = \quad 0.9 \times 100. \tag{C.1}$$

For each of these 90 heads, the outcome of each of the corresponding 90 second flips does not depend on of the outcome of the first flip, so we would expect

$$81 \quad = \quad 0.9 \times 90, \tag{C.2}$$

of these 90 second flip outcomes to be heads. In other words, 81 out of 100 pairs of coin flips should yield two heads. The figure of 90 heads was obtained from Equation C.1, so we can rewrite Equation C.2 as

$$81 \quad = \quad 0.9 \times (0.9 \times 100) = 0.81 \times 100, \tag{C.3}$$

where 0.9 is the probability $p(x_h)$ of a head, so the probability of obtaining two heads is $p(x_h)^2 = 0.9^2 = 0.81$.

A similar logic can be applied to find the probability of the other pairs $(x_h, x_t)$ and $(x_t, x_t)$. For the pair $(x_t, x_t)$, there are (on average) 10 tails observed in the set of 100 first flip outcomes. For each of these 10 flips, each of the corresponding 10 second flips also has an outcome, and we would expect $1 = 0.1 \times 10$ of these to be a tail too, so that one out of 100 pairs of coin flips should consist of two tails $(x_t, x_t)$.

The final pair is a little more tricky, but only a little. For the ordered pair $(x_h, x_t)$, there are (on average) 90 heads from the set of 100 first flips, and we would expect $9 = 0.1 \times 90$ of the corresponding 90 second flips to yield a tail, so nine out of 100 pairs of coin flips should be $(x_h, x_t)$ tails. Similarly, for the ordered pair $(x_t, x_h)$, there are (on average) 10 heads in the set of 100 first flips, and we would expect $9 = 0.1 \times 90$ of the corresponding nine second flips to yield a tail, so nine out of 100 pairs of coin flips should be $(x_t, x_h)$. If we now consider the number of pairs that contain a head and a tail *in any order* then we would expect there to be $18 = 9 + 9$ pairs that contain a head and a tail. Notice that the figure of 90 heads was obtained from $90 = 0.9 \times 100$, so we can write this as $9 = (0.1 \times 0.9) \times 100$, or $p(x_h)p(x_t) \times 100$.

In summary, given a coin that lands heads up on 90% of flips, in any given pair of coin flips we have (without actually flipping a single coin) worked out that there is an 0.81 probability of obtaining two heads, an 0.01 probability of obtaining two tails, and an 0.18 probability of obtaining a head and a tail. Notice that these three probabilities sum to one, as they should. More importantly, the probability of obtaining each pair of outcomes is obtained by multiplying the probability associated with each individual coin flip outcome.

**Conditional Probability**

The *conditional probability* $p(x|y)$ that $X = x$ *given that* $Y = y$

$$p(x|y) \;=\; p(x, y)/p(y), \tag{C.4}$$

where the vertical bar is read as *given that*.

**The Product Rule**

Multiplying both sides of Equation C.4 by $p(y)$ yields the *product rule*

$$p(x, y) \;=\; p(x|y)p(y). \tag{C.5}$$

**The Sum Rule and Marginalisation**

The *sum rule* is also known as the law of total probability. In the case of a discrete variable,

$$p(x) \;=\; \sum_i p(x, y_i), \tag{C.6}$$

and applying the product rule yields

$$p(x) \;=\; \sum_i p(x|y_i)p(y_i). \tag{C.7}$$

In the case of a continuous variable, the sum and product rules yield

$$p(x) = \int_y p(x, y)\, dy = \int_y p(x|y)p(y)\, dy. \qquad \text{(C.8)}$$

This is known as *marginalisation*, and yields the marginal probability $p(x)$ of the joint probability distribution $p(X, Y)$ at $X = x$.

## Bayes' Rule

If we swap $y$ for $x$ in Equation C.5 then

$$p(y, x) \;=\; p(y|x)p(x), \qquad \text{(C.9)}$$

where $p(y, x) = p(x, y)$. Therefore,

$$p(y|x)p(x) \;=\; p(x|y)p(y). \qquad \text{(C.10)}$$

Dividing both sides of Equation C.10 by $p(x)$ yields *Bayes' rule*[26] (which is also known as *Bayes' theorem*),

$$p(y|x) \;=\; \frac{p(x|y)p(y)}{p(x)}. \qquad \text{(C.11)}$$

Within the Bayesian framework, $p(y|x)$ is called the *posterior probability*, $p(x|y)$ is the *likelihood*, $p(y)$ is the *prior probability*, and $p(x)$ is the *marginal likelihood*. Given that this is true for every individual value, Bayes' rule must also be true for distributions of values, so that

$$p(Y|X) \;=\; \frac{p(X|Y)p(Y)}{p(X)}, \qquad \text{(C.12)}$$

where $p(Y|X)$ is a family of posterior distributions (one distribution per value of $x$), $p(X|Y)$ is the corresponding family of likelihood functions, $p(X)$ is the marginal likelihood distribution, and $p(Y)$ is the prior distribution of $Y$.

A brief introduction to Bayes' rule can be downloaded from here: https://jamesstone.sites.sheffield.ac.uk/books/bayes-rule.

# Appendix D

# Key Equations

**Entropy**

$$H(X) = \sum_{i=1}^{m} p(x_i) \log \frac{1}{p(x_i)} \tag{D.1}$$

$$H(X) = \int_x p(x) \log \frac{1}{p(x)} \, dx \tag{D.2}$$

**Joint entropy**

$$H(X,Y) = \sum_{i=1}^{m} \sum_{j=1}^{m} p(x_i, y_j) \log \frac{1}{p(x_i, y_j)} \tag{D.3}$$

$$H(X,Y) = \int_x \int_y p(x,y) \log \frac{1}{p(x,y)} \, dy \, dx \tag{D.4}$$

$$H(X,Y) = I(X,Y) + H(X|Y) + H(Y|X) \tag{D.5}$$

**Conditional Entropy**

$$H(Y|X) = \sum_{i=1}^{m} \sum_{j=1}^{m} p(x_i, y_j) \log \frac{1}{p(y_j|x_i)} \tag{D.6}$$

$$H(X|Y) = \sum_{i=1}^{m} \sum_{j=1}^{m} p(x_i, y_j) \log \frac{1}{p(x_i|y_j)} \tag{D.7}$$

$$H(X|Y) = \int_y \int_x p(x,y) \log \frac{1}{p(x|y)} \, dx \, dy \tag{D.8}$$

$$H(Y|X) = \int_y \int_x p(x,y) \log \frac{1}{p(y|x)} \, dx \, dy \tag{D.9}$$

$$H(X|Y) = H(X,Y) - H(Y) \tag{D.10}$$

$$H(Y|X) = H(X,Y) - H(X) \tag{D.11}$$

*Key Equations*

From which we obtain the *chain rule for entropy*

$$H(X,Y) = H(X) + H(Y|X) = H(Y) + H(X|Y) \qquad \text{(D.12)}$$

## Mutual Information

$$I(X,Y) = \sum_{i=1}^{m}\sum_{j=1}^{m} p(x_i, y_j) \log \frac{p(x_i, y_j)}{p(x_i)p(y_j)} \qquad \text{(D.13)}$$

$$I(X,Y) = \int_y \int_x p(x,y) \log \frac{p(x,y)}{p(x)p(y)} \, dx \, dy \qquad \text{(D.14)}$$

$$
\begin{aligned}
I(X,Y) &= H(X) + H(Y) - H(X,Y) & \text{(D.15)}\\
&= H(X) - H(X|Y) & \text{(D.16)}\\
&= H(Y) - H(Y|X) & \text{(D.17)}\\
&= H(X,Y) - [H(X|Y) + H(Y|X)] & \text{(D.18)}
\end{aligned}
$$

## Channel Capacity

$$C = \max_{p(X)} I(X,Y) \text{ bits} \qquad \text{(D.19)}$$

If the channel input $X$ has variance $P$, the noise $\eta$ has variance $N$, and both $X$ and $\eta$ are iid Gaussian variables then $I(X,Y) = C$, where

$$C = \frac{1}{2} \log \left(1 + \frac{P}{N}\right) \text{ bits,} \qquad \text{(D.20)}$$

where the ratio of variances $P/N$ is the signal-to-noise ratio.

## Marginalisation

$$p(x_i) = \sum_{j=1}^{m} p(x_i, y_j), \qquad p(y_j) = \sum_{i=1}^{m} p(x_i, y_j) \qquad \text{(D.21)}$$

$$p(x) = \int_y p(x,y) \, dy, \qquad p(y) = \int_x p(x,y) \, dx$$

$$\text{(D.22)}$$

## Rate Distortion Theory

Rate distortion theory defines the *rate distortion function* $R(D)$ as the smallest number of bits or rate $R$ that must be allocated to represent $X$ as a distorted version $\hat{X}$ for a given level of distortion $D$,

$$R(D) = \min_{p(\hat{x}|x)} I(X; \hat{X}) \text{ bits} \qquad \text{(D.23)}$$

# Bibliography

[1] Applebaum, D. (2008). *Probability and Information An Integrated Approach, 2nd Edition.* Cambridge University Press, Cambridge.

[2] Avery, J. (2012). *Information Theory and Evolution.* World Scientific Publishing.

[3] Baeyer, H. (2005). *Information: The New Language of Science.* Harvard University Press.

[4] Bialek, W. (2012). *Biophysics: Searching for Principles.* Princeton University Press.

[5] Bishop, C. (2006). *Pattern Recognition and Machine Learning.* Springer.

[6] Cover, T. and Thomas, J. (1991). *Elements of Information Theory.* New York, John Wiley and Sons.

[7] Feynman, R., Leighton, R., and Sands, M. (1964). *The Feynman Lectures on Physics.* Basic Books.

[8] Friston, K. (2010). The free-energy principle: a unifed brain theory? *Nature Review Neuroscience,* 11(2):127–138.

[9] Gatenby, R. and Frieden, B. (2013). The critical roles of information and nonequilibrium thermodynamics in evolution of living systems. *Bulletin of Mathematical Biology,* 75(4):589–601.

[10] Gibson, J. (2014). Information theory and rate distortion theory for communications and compression. *Synthesis Lectures on Communications,* 6(1):1–127.

[11] Gleick, J. (2012). *The Information.* Vintage.

[12] Guizzo, E. (2003). The essential message: Claude Shannon and the making of information theory. *http://dspace.mit.edu/bitstream/handle/1721.1/39429/54526133.pdf. Massachusetts Institute of Technology.*

[13] Hawking, S. (1975). Particle creation by black holes. *Communications in Mathematical Physics,* 43(3):199–220.

[14] Jaynes, E. and Bretthorst, G. (2003). *Probability Theory: The Logic of Science.* Cambridge University Press, Cambridge.

[15] Landauer, R. (1961). Irreversibility and heat generation in the computing process. *IBM Journal of Research and Development*, 5:183–191.

[16] MacKay, D. (2003). *Information Theory, Inference, and Learning Algorithms.* Cambridge University Press.

[17] Nemenman, I., Shafee, F., and Bialek, W. (2002). Entropy and inference, revisited. In Dietterich, T., Becker, S., and Ghahramani, Z., editors, *NIPS 14.* MIT Press.

[18] Paninski, L. (2003). Estimation of entropy and mutual information. *Neural Computation*, 15(6):1191–1253.

[19] Pierce, J. (1980). *An introduction to information theory: Symbols, signals and noise.* Dover.

[20] Reza, F. (1961). *Information Theory.* New York, McGraw-Hill.

[21] Rieke, F., Warland, D., de Ruyter van Steveninck, R., and Bialek, W. (1997). *Spikes: Exploring the Neural Code.* MIT Press, Cambridge, MA.

[22] Seife, C. (2007). *Decoding the Universe: How the New Science of Information Is Explaining Everything in the Cosmos, From Our Brains to Black Holes.* Penguin.

[23] Shannon, C. (1948). A mathematical theory of communication. *Bell System Technical Journal*, 27:379–423.

[24] Shannon, C. (1959). Coding theorems for a discrete source with a fidelity criterion. *IRE Nat. Conv. Rec*, 4(142-163):1.

[25] Shannon, C. and Weaver, W. (1949). *The Mathematical Theory of Communication.* University of Illinois Press.

[26] Stone, J. (2013). *Bayes' Rule: A Tutorial Introduction to Bayesian Analysis.* Sebtel Press.

[27] Stone, J. (2022). *Information Theory: A Tutorial Introduction (Second Edition).*

[28] Taylor, S., Tishby, N., and Bialek, W. (2007). Information and fitness. *arXiv preprint arXiv:0712.4382.*

[29] Tkačik, G. and Bialek, W. (2016). Information processing in living systems. *Annual Review of Condensed Matter Physics*, 7:89–117.

[30] Wallis, K. (2006). A note on the calculation of entropy from histograms. Technical report, University of Warwick, UK.

# Index

alphabet, 77
average, 77

bandwidth, 66
Bayes' rule, 85, 88
binary
    digits vs bits, 16
    number, 77
binary rate distortion function, 70
bit, 13, 18, 33, 77
byte, 10, 77

capacity, 45, 77, 90
chain rule for entropy, 90
channel, 27, 57, 77
channel capacity, 4, 45, 46, 77
code, 77
codebook, 63, 77
codeword, 77
communication channel, 4
conditional
    entropy, 78
    probability, 78, 87
continuous, 78

data processing inequality, 62
die, 22
difference coding, 7
differential entropy, 31, 65, 78
discrete variable, 78
distortion, 69

encoding, 4, 38, 57, 78
entropy, 18, 21, 78
    conditional, 78
    definition, 21
    differential, 31
    exponential distribution, 41
    Gaussian distribution, 42
    maximum, 37
    negative, 41
    uniform distribution, 39
entropy vs information, 26
equivocation, 69, 78
expected value, 78
exponential distribution, 41

fly, 38

Gaussian
    channel, 64
Gaussian distribution, 42
Gaussian rate distortion function, 71

histogram, 27, 78
Huffman coding, 38

iid, 78
independence, 6, 13, 78
information, 16, 26, 33, 78
information theory, 4
information vs entropy, 26

law of large numbers, 48, 78
logarithm, 15, 18, 79

marginalisation, 88, 90

www.ingramcontent.com/pod-product-compliance
Lightning Source LLC
LaVergne TN
LVHW050144060326
832904LV00004B/172

* 9 7 8 1 7 3 9 6 7 2 7 2 0 *